THE
MAKING OF
THE EARTH

A new**scientist** GUIDE

THE MAKING OF THE EARTH

Edited by
RICHARD FIFIELD

Basil Blackwell & New Scientist

© Articles and editorial, IPC Magazines Ltd.
Volume rights, Basil Blackwell Limited, 1985.

First published in book form in 1985 by
Basil Blackwell Limited
108 Cowley Road, Oxford OX4 1JF.

Basil Blackwell Inc.
432 Park Avenue South, Suite 1505
New York, NY 10016, USA.

British Library Cataloguing in Publication Data
Fifield, Richard
    The making of the earth——(New scientist guides)
    1. Plate tectonics
    I. Title   II. Series
    551.1'36   QE511.4

    ISBN 0-631-14237-1
    ISBN 0-631-14238-X Pbk

Library of Congress Cataloging in Publication Data
The making of the earth——(New scientist guides)
Includes index.
    1. Earth sciences——Addresses, sciences——Addresses,
    essays, lectures.
    I. Fifield, Richard.   II. Series: New scientist guide.
    QE35.M215   1984   550   84-28435

    ISBN 0-631-14237-1
    ISBN 0-631-14238-X (pbk.)

Typeset by Katerprint Co Ltd, Oxford
Printed and bound in Great Britain by
T. J. Press, Padstow

Contents

Contributors

CHESTER BEATY is professor of geography at the University of Lethbridge in Alberta, Canada.

DR CLIVE BISHOP is keeper of mineralogy at the Natural History Museum in London.

DR BASIL BOOTH, formerly a visiting professor at Iowa State University, is now a consultant vulcanologist and runs a scientific photographic library.

DR HOWARD BRABYN is English editor of the *Unesco Courier* based in Paris.

JAMES BRANDER wrote his articles whilst a research assistant in the department of geophysics at Imperial College, London.

PROFESSOR JOE CANN is head of the department of geology at the University of Newcastle-upon-Tyne.

PRESTON CLOUD is professor of geology at the University of California in Santa Barbara.

MARGARET EVANS is a freelance writer based in British Columbia.

RICHARD FIFIELD has been managing editor of *New Scientist* since 1969.

PROFESSOR DAVID FISHER is with the Institute of Marine Sciences at the University of Miami.

DR R. L. FLEISCHER wrote on fossil records of nuclear fission whilst working with Professor P. B. Price and Professor R. M. Walker at the General Electric Research Laboratory in Schenectady, New York.

DR PETER FRANCIS lectures for the department of earth sciences at the Open University.

LAURA GARWIN wrote on fission track dating while a research fellow in the department of earth sciences at the University of Cambridge.

DR TOM GASKELL wrote about prospecting for oil in the North Sea when he was an employee of British Petroleum Limited.

DR ROLAND GOLDRING is a lecturer in geology at Reading University.

DR RICHARD A. F. GRIÈVE has recently left the department of geological sciences at Brown University, Rhode Island, and is now with the earth physics branch of Energy, Mines and Resources in Ottawa.

DR HANS HASS wrote his article "A new theory of atoll formation" whilst doing research for the Internationales Institut für Submarine Forschung in Vaduz, Liechtenstein. After making a television series based on his research in the Maldives, he subsequently wrote a book *Journey into the Unknown* published by Hutchinsons.

DR ANN HENDERSON-SELLERS is a lecturer in the department of climatology at the University of Liverpool.

DR JAMES JACKSON was co-director of the seismology team of the Royal Geographic Society's International Karakorum Project and is currently a research fellow at Queen's College, Cambridge.

DR GEORGES LACLAVÈRE wrote on the Upper Mantle Project whilst at the International Union of Geodesy and Geophysics based at the Observatoire Royal in Brussels. He now resides in Paris.

DR B. J. LEVIN is at the O. Y. Schmidt Institute of Physics of the Earth, affiliated to the Academy of Sciences of the USSR in Moscow.

SEAN MCCUTCHEON is a freelance writer based in Montreal.

PROFESSOR D. F. MERRIAM is chairman of the department of geology at Wichita State University in Kansas. He wrote about geology and computers whilst at Leicester University.

DR STEPHEN MOORBATH, FRS, is with the department of geology and mineralogy at Oxford University.

DR ROBERT MUIR WOOD was a member of the Royal Geographic Society's International Karakorum Project. He is now at Trinity College, Cambridge.

DR JOHN NORMAN AND MUO CHUKWU-IKE wrote about the implications of crustal fractures whilst at Imperial College, London.

PROFESSOR P. B. PRICE is currently with the department of physics at the University of California at Berkeley.

PROFESSOR A. E. RINGWOOD, FRS, is with the school of earth sciences at the Australian National University in Canberra.

CONSTANTIN ROMAN wrote his article "Buffer plates: where continents collide" whilst researching with the department of geodesy and geophysics at Peterhouse, Cambridge.

PROFESSOR A. E. SCHEIDEGGER wrote on the origin of continents when he held an associate professorship of mathematics at the University of Alberta, Canada.

STEPHEN SELF is professor of geology at the University of Texas.

STEPHEN SPARKS wrote about the Heimaey eruption during the time he was a research student of vulcanology with Stephen Self at Imperial College, London.

DR PETER STUBBS is a former deputy editor of *New Scientist*.

DR CHRISTINE SUTTON is a freelance writer and former physical sciences editor of *New Scientist*.

PROFESSOR L. R. WAGER wrote about measuring geological time when he was at the Department of Geology and Mineralogy, University Museum, Oxford.

PROFESSOR R. M. WALKER is director of the McDonnel Center for Space Sciences at Washington University, Missouri.

DR TONY WATTS is at the Lamont-Doherty Geological Observatory of Columbia University, New York.

DR J. TUZO WILSON OBE, FRS, is currently Director General of the Ontario Sciences Centre.

DR ALAN WOOLLEY is at the department of mineralogy at the Natural History Museum in London.

Foreword

In November 1956 a small and enthusiastic team of journalists and scientists launched *New Scientist*. Ever since then, with a few interruptions, *New Scientist* has provided a weekly dose of news from the world of science and technology.

Over the years, the magazine has reported research as it happened. Sometimes the findings that we described turned out to be less enduring than their discoverers first thought. Sometimes they went on to win Nobel prizes for the scientists involved. Theories that first provoked severe scepticism became established wisdom. Established wisdom was cast aside.

These guides bring together that 'history in the making'. In these pages you will find more than a scientific account of a particular subject. You will find the personalities and the problems; the excitement of discovery and the disappointment of wrong turnings and the frustration of delays. I hope, too, that you will find something of the excitement that scientists feel as they push back the frontiers of knowledge.

Science writing does not stand still any more than does science. So the past quarter of a century has seen changes in the way in which *New Scientist* has presented its message. You will find those changes reflected here. Thus this guide is more than a collection of articles carefully plucked from the many millions of words that have appeared over the years. It is a record of science in action.

Michael Kenward
Editor, New Scientist

Preface

The eminent physicist P. M. S. Blackett commented in a lecture to the Royal Institution of Great Britain, on Friday 23 March, 1956, that "Clegg, Almond and Stubbs were the first to prove definitely by this means [fossil magnetism] the movement of land masses relative to the poles".

This seemingly unimportant statement marked a high point in one of the most significant developments in the Earth sciences. The scientific revolution of which fossil magnetism is a part has roots that go back to the early decades of the 20th century, and some even older ones. The suggestion that the continents had split apart from each other in geological history had been around for some centuries, but the claim that the continents were not fixed but could wander about seemed altogether too bizarre even to many open-minded geologists in the early decades of the 20th century. Like all scientific revolutions, much of the scientific establishment was against "continental drift" because no one could explain what caused it.

At the beginning of the 17th century, the English essayist and philospher Francis Bacon had noticed that the outline of the eastern side of the Americas and the western side of Africa looked so similar that it seemed possible that they were once joined together. With the new sense of exploration and adventure in the centuries that followed, and especially as means of travel and communication improved, news filtered back of huge coal deposits on the European and American shores of the Atlantic. The geologists were quick to notice that the fossil plants that turned up in the American and European coals were often of the same species.

In 1859, the American geologist Antonio Snider-Pellegrini wrote a book which he ambitiously entitled *La Création et ses*

Mystères dévoilés. He included in his book two maps which he claimed showed the alteration of the relative positions of land and water since the creation. Snider believed in the Noachian deluge. He supposed the Earth once to have consisted of a continuous block, or mass, rising from the ocean, and that the mighty Atlantic Ocean had formerly been dry land. The two continents were formed by the division of the former block, and he believed the coals had formed from the materials washed from the continents as they divided.

A new enthusiasm for the idea that the continents had moved laterally came in the writings of the American geologist F. B. Taylor (in 1910) and the German meteorologist Alfred Wegener (in 1912). Wegener saw the similarity of the fit of the coastlines of Brazil and Africa as providing the "starting point of a new conception of the nature of the Earth's crust and of the movements occurring therein; this new idea is called the theory of the displacement of continents, or, more shortly, the displacement theory, since its most prominent feature is the assumption of great horizontal drifting movements which the continental blocks underwent in the course of geological time, and which presumably continue even today".

Within the past 25 years a new theory has revolutionised our understanding of the Earth's crust and the formation of the continents. The theory argues that the Earth's surface is made up of some 15 quite rigid "plates" in continuous motion relative to each other. Where they are moving apart in mid-ocean, ridges form. Where they converge, one plunges down into the mantle beneath, and ocean trenches are formed. Where an ocean plate meets continental crust, mountain ranges tend to be formed, as with the Andes. The interaction between the edges of the plates accounts for almost all earthquake activity. This theory sees the ocean beds as being continuously transformed. They are relatively recent compared with the rocks that constitute the basements of the great landmasses. The importance of the theory of "plate tectonics", as it has come to be known, is that it offers not only a physical basis for the ideas of continental drift, but that it provides a verifiable theory about the nature and history of the Earth's surface and of its inner regions.

Professor Blackett's discourse in 1956 was significant not just because an eminent physicist publicly gave credit to his colleagues and their ideas, but because it showed how valuable it is that scientists from various disciplines should work on a major topic.

Geophysics, of which fossil magnetism ("palaeomagnetism") had become a part, was in the years ahead to provide new ways of viewing and investigating our planet.

The Soviet Union launched the first man-made Earth satellite, Sputnik, on 4 October, 1957. The achievements of that satellite would contribute to the successes of the International Geophysical Years, for which plans had been drawn up in 1956. The IGYs were to become the most significant international cooperative venture in science. In turn, the IGYs would spawn the Antarctic Treaty and greatly enhance professional and public interest in "spaceship Earth". Sputniks became regular features of the night skies.

New Scientist was launched on 22 November, 1956, for a public eager for scientific information.

The new science journal and its readership were fortunate in that Peter Stubbs, one of the colleagues to whom Blackett referred in his RI lecture, was enticed away from his research to explain just what geophysics was all about. Soon, Stubbs joined the editorial staff of *New Scientist*, thus putting the journal in a unique position — unique for what was considered by its proprietors to be a newspaper of science and technology. Stubbs was to become not only one of the journal's most eminent physical science editors but also later its deputy editor.

Peter Stubbs had a considerable flair for communicating science and he had an enviable style of writing. He remained with *New Scientist* for over 20 years. That period saw the victory of the revolution in the Earth sciences against the establishment diehards, and it also saw the realisation of the implications of the revolution and the first harvest of fruits. *New Scientist* was able to chronicle the developments of the revolution leading to the testifiable theory of plate tectonics. As Professor Tuzo Wilson explains in his introduction (an article that he wrote to celebrate the journal's 25th anniversary): "As befits a mighty subject, it has needed many Lilliputians to assemble our fascinating picture of a mobile Earth."

Many strands of the complex story of the Earth's history have been successfully untied, but a multitude of knots have yet to be untangled. Even so, as Professor Preston Cloud emphasises in his chapter, we have enough threads to weave "a strong, if coarse-textured, tapestry of events". This book offers a view of that tapestry as presented in the pages of *New Scientist*.

For convenience, I have organised my selection of articles and news pieces into four parts.

The first part is concerned to set the clock back, and to look at some of the detective work that was required to unravel aspects of the Earth's long history, to consider how the planet might have been formed, and how it gained an atmosphere that could support life.

In the second part, I have included articles that provide support to the theories of continental drift, sea-floor spreading, and the concomitant theory of plate tectonics.

Part three looks at some of the insights that emerge if we accept the idea that the oceans and continents are carried on mobile, lithospheric plates that sometimes collide and override each other.

The final part hints at some of the pay-offs that have come from various geophysical techniques developed over the past few decades and from our understanding of plate tectonics. That final section also indicates some of the things that we can expect to emerge from our new-found ability to look back at planet Earth from space.

This compilation would not have been possible without the painstaking and dedicated work of *New Scientist*'s book researcher, Jane Moore. She compiled a list of the journal's entire coverage of the Earth sciences, for which I am sincerely grateful. I would also like to thank Karen Iddon, my secretary, for her assistance and patience which enabled me to complete my task of producing a final manuscript. As always, Neil Hyslop shouldered the task of preparing the original line drawings.

Richard Fifield

PART ONE

Introduction

1

Geology: a historical perspective

RICHARD FIFIELD

Early interest in the Earth and its rocks can be traced back to the ancient Greeks, including Herodotus, Pythagoras and Strabo. The Roman Seneco and the Chinese Zhang Heng were both interested in the causes of earthquakes, and Heng (AD 78–139) invented the first seismograph with which to record an earthquake. Avicenna (980–1037), the Persian physician and philosopher, became the father of the Earth sciences in the Arab world. After the Renaissance, Leonardo da Vinci (1452–1519) correctly explained that fossils were remains of former living creatures. Georg Agricola (1494–1555) did much to rationalise aspects of metalliferous mining with his *De re Metallica*, published in Basle in 1556.

The strict application of the Old Testament dominated thinking from the 16th through to the middle of the 19th century, with beliefs in the Biblical Flood of Noah's time causing denudation, and subsequent erosion, earthquakes and volcanoes responsible for decay. Bishop James Ussher (1581–1656) calculated from the Old Testament that the Earth was created in the year 4004 BC. The French naturalist Comte Georges Buffon (1707–88) deduced that the age of the Earth must run to millions of years. Scotsman James Hutton (1726–97) argued that the Earth's antiquity was unfathomable: "we find no vestige of a beginning, no prospect of an end".

Hutton became the founder of "modern geology". He developed theories about Earth processes. He accepted the idea of denudation and considered the topography of the continents to be sculptured by rain and river action. Hutton explained concepts of sedimentation and lithification, and the formation of strata. He claimed that from time to time rocks were raised from the sea to form new continents. His ideas are summarised as the *principle of uniformity* – the present is the key to the past: observation of Earth processes that are operating today enable us to interpret the various products of those same processes that operated in earlier times. Hutton's countryman Charles Lyell developed and expanded on those views in his famous book *Principles of Geology* (1830–33).

The increasing pace of industrialisation from the end of the 18th century led to widespread development of systems of canals and roads, and later the building of vast railways. The work involved in constructing these systems of transport brought to light much new geological information. The importance of knowledge of the rocks for such work to building transport systems meant that geology could no longer be left to the military, the miner or the interested amateur as hitherto it had been. The first half of the 19th century witnessed an increasing professionalism which was gradually matched by the development of better-organised geological societies, geological surveys and geological map departments.

The Oxfordshire-born William Smith, the father of stratigraphy and English geology, discovered in the 1790s that he could distinguish various strata by the fossils that they contained, and he used this discovery to help him to construct the first geological map of England and Wales, which was published in 1815. The techniques of applied palaeontology proved invaluable in working out geological structures and was taken up in all other countries. It became possible to sort out the stratigraphy of the Phanerozoic era (the period from 580 million years ago to today, during which life became abundant on Earth). Other techniques were required for rocks that were igneous in origin and contained no fossils, or which had become changed (metamorphosed) beyond recognition by heat and the pressures of compaction and Earth movements. Here the experience of the mineralogists whose special interests were the mineral and metalliferous components of rocks, came into their own. In the early 19th century the mineralogists were greatly aided in their work by the Scottish physicist William Nicol who invented the polarising prism. Fitted into a microscope, these prisms revolutionised the analysis of rocks: the petrographic, or "polarising", microscope could easily be taken into the field by the geologist. Igneous rocks could be classified by their mineralogical content into specific types. Even if the igneous rocks could not be "dated", it was often possible to draw inferences on their age by their relationship with other rocks containing fossils.

From the 19th century onwards, research and improvements in surveying techniques led to numerous ways of assessing geological structures. Through the work of George Everest (1790–1866) and John Pratt (1809–71) in the Himalayas, and of the astronomer George Airy (1801–92), it became clear that mountains have roots. The mass of a mountain or the mass of a large submerged metal deposit should deflect a sensitive pendulum. Pratt found, however, that mountains invariably failed to deflect a sensitive pendulum to the extent that theory indicated their mass should. Airy conceived the idea that mountains and continents rest on a much denser base. As mountains are eroded, or snow melts from a continent, the land rises and the settling sediments compensate by depressing some other parts of the Earth. The American geologist Clarence Dutton (1841–1912) proposed the term "isostasy" for this effect. The techniques of "gravity surveying" emerged from such ideas.

From the earliest days of research on radioactivity, chemists and physicists realised that it could provide a means of dating rocks. Radioactive elements decay at a specific rate, regardless of temperature or pressure. Thus geologists can gauge the absolute age of a rock by comparing in that rock the proportions of radioactive elements and the daughter products into which they decay. It was some decades, however, before sufficiently accurate and reliable technology was available to cope with the tasks involved.

During the 20th century, a host of new techniques has emerged for prospecting and analysis. Seismic reflection, electrical conductivity, magnetic measurement and the reflection of ultrasound (radiosonde), have all provided ways of profiling rock sequences, and have been widely used in the hunt for oil and mineral deposits. These techniques, together with many sophisticated forms of remote sensing from aeroplanes and satellites, and computerised image analysis, add to the armaments of the modern geologist.

It was work on cosmic rays, however, after the Second World War, that led P. M. S. Blackett (1897–1974) and his team at Manchester University to build a highly accurate magnetometer. During that work, the team measured the residual magnetism in some Triassic sandstones. These findings showed that in the past 150 million years part of Britain has drifted northwards from a position much nearer the Equator than it is at present.

The revelation that the continents and oceans are actually carried on mobile plate-like components of the Earth's crust emerged from a welter of evidence collected during the 1960s and 1970s. Now, in mid-1984, scientists from the US National Aeronautics and Space Administration have released the first accurate measurements of the plate movements. They obtained the data using laser and satellite techniques, and showed that the plates are moving at rates between 1.5 and 7 centimetres per year at different parts of the globe.

The Earth sciences have now reached a new "normal phase" in a development that the philosopher Thomas Kuhn claims is typical of all scientific revolutions. Few Earth scientists now question whether plate tectonics is "right". The debate is now whether the configuration of the continents at various times in the past was this way or that way; whether the convection mechanism for driving the plates can be worked out in detail; whether fossil evidence indicates that Antarctica and India were once adjacent; and so on. Science, like the continents, is always on the move.

2

Movements in Earth science

J. TUZO WILSON

In the International Geophysical Year of 1956 scientists produced the first map to show the global mid-ocean ridge system; a decade or so later sea-floor spreading at the ridge became part and parcel of the theory of plate tectonics, thus vindicating the concept of continental drift. Here, a geophysicist who witnessed more than 50 years of the development of this story describes some of the characters who took part in it.

It is now 58 years ago that I first, as a schoolboy, spent a summer in the field, and nearly as long since I entered geophysics. There are few – except three of my mentors who, I am glad to say, at 90 or so are still alive and well – who have first-hand knowledge over so long a time. Probably there is no one else who can say that it was his fate to try to study geophysics at three first-rate universities, all of which were trying to begin the teaching of that subject, but none of which had yet found out how to do it. That earlier time is what I intend to emphasise and I must further confine my words to the solid Earth.

It was in 1908 that I was born in Ottawa, Canada, first child of a happy but serious-minded couple who 15 summers later dispatched me to the northern woods as a schoolboy assistant on a forestry party. For three of the next five seasons I worked as field assistant to Noel Odell, today still alive and well in Cambridge. At that time he had just come back from Mount Everest where he had gone as a geologist and returned a hero. Naturally his reputation and his kindly manner made a profound impression on me. Although I had in the meantime entered the University of Toronto and passed the first year in physics successfully, I decided that I should prefer a life in the woods unravelling the mysteries of geology to one in the lab where the practice of physics, then in its heyday, seemed repetitious and stuffy, much as I admired the elegance of its theories.

In the autumn of 1927 when I proposed this transfer all the university authorities were shocked. Professor, later Sir, John McLennan was dismayed and irritated that any promising student should abandon so prestigious a subject as physics for geology, then held in very low regard. Nor did the geologists, who at that time only mapped little patches and identified the rocks and fossils which they encountered, want anyone who might expose their limitations. To their general hostility there was fortunately one exception.

During the First World War many physicists had been employed as artillery sound-rangers and they could not help but notice the existence of ground waves, which indeed Ludger Mintrop had already begun to study in Germany about 1908. When the war was over, John C. Karcher, whom I had the good fortune to meet later, investigated these waves which he generated by small explosions, and in 1921 he obtained a patent on the seismic method of prospecting for petroleum. This was immediately successful and led to a great surge in activity and incidentally to the revival of electrical and magnetic methods first used in Cornwall and in Sweden respectively more than a century earlier.

When the first geophysical prospectors descended upon mining camps no one there knew what to make of them and they could not tell which operators were genuine and which were charlatans. All were regarded in the same class as rain-makers. Soon the United States Bureau of Mines and the Geological Survey of Canada decided to investigate and they employed Professors A. S. Eve and D. A. Keys of McGill University, and Lauchlan Gilchrist of Toronto to study and test the methods. By the time I asked for a transfer, this work was well begun and Gilchrist had just decided to start a course in mining prospecting and was on the lookout for students. He took me under his wing, arranged that I take a double major in physics and geology, but he was able to arrange for only two very incomplete lecture courses in geophysics in one of which I was the only student. Nevertheless, when I graduated in 1930 I could claim that I had by accident become the first to graduate in Canada in a course designed to train geophysicists. Of course there were practising geophysicists already, but they had picked up their expertise on the job. In total there was, so far as I knew, one man who measured gravity, one or perhaps two seismologists, four or five people studying the Earth's magnetic field, two physical oceanographers, a few meteorologists and geodetic surveyors and a rapidly increasing number of geophysical prospectors, some very able and some complete mountebanks.

As no one quite knew what geophysics was and as it was then very easy to get good marks in geology I had the good fortune to win a scholarship to Cambridge where I naturally went to study under Sir Harold Jeffreys at St John's College. He also is still alive and well, a great figure of whom the late Sir Edward Bullard (no slouch himself) once said to me, "The only time I feel in the presence of genius is when I go to see Harold." I dutifully attended Jeffreys's eight lectures. Unfortunately I could neither hear nor understand them, but Jeffreys was accustomed to this and did not hold it against me. Nevertheless, I was more at home with my tutor, Sir James Wordie. Although a very astute Scot, his chief claim to fame was that he had wintered in the Antarctic, on one of Shackleton's less successful expeditions, under an overturned lifeboat on a diet of raw seal meat and penguins. I saw him quite often, not always for the best of reasons, but he patiently interested me in the joys of exploration, and did not seem upset that I spent my two years in Cambridge in travelling, rowing, flying and drinking. At least those are the parts I remember.

Looking back, I suppose that Wordie was vaguely guilt-ridden for encouraging me to come to Cambridge under mildly false pretences. Certainly as a student and ever afterwards he treated me with the greatest kindness.

The situation as far as I can find out was something like this. A very able mathematician and engineer, Sir Gerald Lenox-Conyngham, when he retired as director of the Survey of India in 1921 realised that detailed surveying needed more well-trained recruits and demanded a better understanding of the Earth. He was able to get a fellowship at Trinity College and a position as an unpaid lecturer in geodesy. He also realised that to achieve his objective he needed the help of an experimental geophysicist. There being none in sight he discovered that the great physicist, Sir Ernest Rutherford, was in the habit of dining in hall at Trinity on Sunday evenings and he so positioned himself on the next convenient Sunday that he sat next to Rutherford at dinner. He asked him whether he could suggest a possible candidate. Rutherford proposed Edward Bullard, a student in nuclear physics, and later advised Bullard to take the job. At the same time those involved initiated a statute to found a Department of Geodesy and Geophysics, but this was not passed until 1931 when I had been at

Cambridge for a year; and meanwhile Bullard had left to do the field work for his PhD, measuring the gravity field over the East African rift valleys. This was a sensible decision, but meant that I did not meet Bullard at that time, although I do remember his very determined wife rushing in to cram some physics so that she could help "Teddy" on the trip which was also to be their honeymoon.

Bullard's reign in Cambridge

The two years I spent in allegedly studying geophysics in Cambridge passed before the department of Geodesy and Geophysics at Madingley Rise existed. Fifty years later I chanced to be in Cambridge when another statute came to the vote and I was able as a graduate to go to the Senate House and vote to unite geodesy and geophysics with the other earth-science departments. Thus I witnessed the whole span of Bullard's reign at Madingley Rise where he built such a brilliant department. By his creative ability and by his custom of inviting colleagues from all over the world to spend their leave there he exerted a major influence on the development of the subject.

He was witty, entertaining and a rebel. A photograph at the Scripps Institute shows a rear view of him seated on a boulder, clad in nothing but a large sombrero. I recall one chilly October evening when he was Head of the Department of Physics at Toronto he persuaded a mixed party all to take a skinny-dip in a lake in northern Ontario. They did not linger. Few people possess such *joie de vivre* and not many of them are such able scientists.

Bullard claimed that he had inherited his original cast of mind from his grandfathers. One, the brewer of the once well-known Bullard's Ale, ran for parliament and was elected, but barred from taking his seat because of an accusation that he had bribed voters with pints of beer and half-crowns. Ten years later he was allowed to run again and again got elected, but this time for the other party! Bullard's other grandfather lived in the flat valley of the Thames. Tiring of the monotonous view, he had a model of the Matterhorn erected at the foot of his garden complete with tin chamois deer and operating waterfalls. His grandchildren were allowed to inspect it from the house through a telescope.

When I returned to Canada at the height of the depression I could find no regular work. The director of the Geological Survey told me that, although he would like to employ one geophysicist, it was impossible and he advised me to take a PhD in geology after which he might find me a position. I accordingly wrote to Harvard University, the Massachusetts Institute of Technology and Princeton University, all of which admitted me, but I chose the last for three good reasons. Princeton was the only one to offer me any money, was the only one that proposed to teach any geophysics and was the only one where I knew anybody. (It's hard now to realise how few then went to university and particularly how few did graduate work.)

Professor R. M. Field who enticed me to Princeton was an extraordinary man. He bubbled with energy, enthusiasm and powers of persuasion. The other professors wished that he had become a salesman, for which he would have been well suited, and they disparaged his ability as a professor. It's true that he did no research, but he held freshmen classes spellbound, he organised great field trips and he had a sweep of imagination lacking in his research-conscious colleagues. In particular, he had been a student at Harvard with Alexander Agassiz, the son of Louis Agassiz, the Swiss geologist and zoologist who had proposed the existence of ice ages and ice sheets. The son was a marine zoologist and inspired Field with the idea that geologists should explore ocean floors.

The only large-scale effort to explore the bottoms of the deeper oceans had been the voyage of HMS *Challenger* and geologists generally supposed that the ocean floors were smooth and gently sloping and that a complete succession of sedimentary strata would overlay an ancient basement.

In 1931 Field heard Maurice Ewing from Lehigh University in Bethlehem, Pennsylvania, give a paper on the theory of seismic prospecting for petroleum at the tiny annual meeting of the American Geophysical Union. Field was smart enough to realise that the same methods which were proving so useful and profitable on land offered the first possibility of mapping the ocean floors in detail and of discovering what lay beneath them.

With his usual powers of persuasion Field got William Bowie, then director of the United States Coast and Geodetic Survey, to produce a grant for $2000 and to accompany him to Lehigh University, where Ewing had a job as an overworked lecturer in physics. Field and Bowie persuaded Ewing to accept the grant and to use it to start the seismic exploration of the world's oceans!

I am sure that Field hoped to bring Ewing to Princeton, but he was unable to persuade the university to put up the money.

Meanwhile Field went about recruiting students and in the autumn of 1933 three of us came to Princeton to study geophysics. D. C. Skeels was a Rhodes scholar from Montana who had degrees in mathematics and in geology. G. P. Woolland was an engineer from Georgia Institute of Technology and I was the third.

Unfortunately Ewing remained at Lehigh, and on a few weekends we drove over to Bethlehem, 160 km away, to help him with his seismic investigations which he wisely began on the coastal plain of New Jersey. There the geology was known and he later extended his work out over the continental shelf to the deep ocean.

At that time Ewing was a happy young man, newly married, very hard working of course, but still willing to take us to his home for supper and a jolly evening chatting. Later he became a compulsive worker and a recluse.

Harry Hess also came to Princeton that year as a lecturer in minerology. He was not yet involved in geophysics. I was fortunate to know these able men, both immensely hard workers, who contributed so much in their very different ways. They were good friends, but perhaps a little jealous of one another. Ewing had had a very hard early life, harder than the biographical story published in the *New Yorker* magazine admitted. This was

Figure 2.1 *The floor of the oceans based on analysis of SEASAT data by William F. Haxby at Lamont-Doherty Geological Observatory, March 1983. (Credit: William F. Haxby/Lamont-Doherty, Geological Observatory)*

perhaps because Ewing edited the story. At any rate he grew up to be a tough leader and one with a great gift for raising money and organising large enterprises. Hess was more of a loner. For several years during the Second World War he commanded a naval transport ship and conducted a one-ship survey of the southwest Pacific Ocean. There he discovered flat-topped seamounts – large underwater volcanic mountains – or guyots as he called them. He once showed me a map he had compiled of the whole Pacific Ocean. But, whereas Hess had completed and published only a few sheets of the Pacific, and some others of the Caribbean, Ewing, Bruce Heezen and Marie Tharp at Columbia University and also Henry Menard, Robert Dietz and Roger Revelle at Scripps had, by organising large teams, compiled and published maps of whole oceans. Nevertheless Hess had geological insights and an under-standing of the Earth that Ewing, who remained a physicist, never achieved, although he devised new instruments and sent out more and more expeditions. Hess put the difference in an exaggerated way when he remarked to me, "No one has collected more data and contributed less ideas towards the study of ocean basins than Maurice." In preparation for the International Geophysical Year of 1956 – the year *New Scientist* was launched – Ewing and Heezen produced the first map to assemble all the known pieces of the mid-ocean ridge system and at the same time suggested that they were all connected into a single, world-embracing feature. They sent out expeditions which four years later had established that this was true.

Indications of the existence of this mid-ocean ridge system had been found by the American oceanographer Matthew Maury in about 1855, and by the *Challenger* and other expeditions. About 1930 Arthur Holmes, professor of geology at Durham University, suggested that there might be upwelling convection currents beneath the ridges. In 1960 Hess, using the new data, extended that idea and established the concept of sea-floor spreading from mid-ocean ridges.

But I have jumped ahead. In the early 1930s neither Hess nor anyone else at Princeton knew much geophysics and as it was impractical to study with Ewing regularly we had to teach ourselves what geophysics we could and take our degrees in geology. My own thesis area in the Beartooth Mountains of Montana introduced me to climbing, great faults, aerial photo-graphs – which were becoming available for the first time – and a west still somewhat wild.

That done, the director of the Geological Survey of Canada gave me that job he had promised and 10 years of mapping and service in the army intervened. At the conclusion of the war the Canadian services made perhaps the only long automotive expedition ever made to the Canadian Arctic. I took a part in Exercise Musk-Ox on which "all-terrain" vehicles travelled 5500 km from the rail-head on Hudson Bay north to the Arctic islands, where one of our party, Morris Innes, relocated the moving north magnetic pole, and thence via Coppermine and the Mackenzie River valley back to rail-head in Alberta. This led to my joining the first United States Air Force flight to the north pole in 1946 and other long flights over the arctic searching for and eventually finding the last unknown islands to be discovered anywhere outside Antarctica.

Returning to civilisation and civilian life I took the good advice of my third 90-year old mentor, C. J. Mackenzie, then President of the National Research Council of Canada, and went to the University of Toronto to do 20 years of research, of the synthesising rather than the analytical type.

It was an exciting period. Suddenly new modern instruments, including mass-spectrometers, flux-gate magnetometers, spinners to measure palaeomagnetism, gravimeters, improved seismographs, instruments to measure heat flow and many more were available together with money to buy and deploy them all over the world on land and sea. The opening of airlines made it relatively cheap, quick and easy to travel all over the globe.

Just as those I have mentioned and others began the exploration of the oceans, so others started to change geology from the mere mapping of many isolated areas into comprehensive studies of whole continents. Determinations of age and mapping of large faults made it possible to divide ancient shield areas into provinces of different ages. In pursuit of this work I was a member of the teams that produced the first tectonic map and the first map of the glacial geology of Canada.

Using aerial photographs as they became available we looked for great faults, but found that over most of Canada they were obscured by glacial drift. Characteristic glacial features showed so plainly that sitting in the office a succession of assistants was able to map the whole pattern of glacial features and latest ice movements which had been largely unknown before. Later I assembled world-wide data on large faults and on ocean islands.

It occurred to many of us, that geology and astronomy differed from other sciences in that the main objects of those studies were too large to bring into the laboratory. In this they contrasted with physiology, chemistry or physics. Nevertheless, if it is sensible to study the physiology of some animal as a whole, so it seemed sensible to study the Earth as a whole, even if that meant first assembling a vast amount of data.

One problem was that whenever one tried to do this one encountered contradictions. Another problem was that whereas other sciences gloried in accurate theories from which one could make reliable predictions, no theories about the Earth seemed to make any sense. This difference between geology and physics had become apparent to me as an undergraduate, but like most others I was too stupid to see the answer to the puzzle.

This was, as the German geophysicist Alfred Wegener had proposed in 1912, that the Earth is actually mobile and that to try to apply theories based upon the idea that it is static make no sense.

A few geologists with great imagination or clear data followed Wegener. These included the Swiss geologist Emile Argand and his colleagues, South African Alexander du Toit and his students, Warren Carey from Tasmania and W. A. J. M. van Watershoot van der Gracht. But as in the case of other revolutions the whole establishment of geological surveys, university departments, mining companies and petroleum organisations opposed any idea of continental drift and it was scarcely mentioned at any of the three universities I attended. New methods gradually produced information of kinds lacking before and these slowly forced most Earth scientists to change their views. A particularly important problem that came to light in the late 1940s and early 1950s was the difficulty of explaining how great displacements on faults could occur on a static Earth.

At the same time, scientists were improving the study of palaeomagnetism and their results indicated continental motion. Victor Vacquier's discovery that the ocean floor is magnetically imprinted in regular patterns and the interpretation of this simultaneously by British geophysicists Fred Vine and D. H. Matthews, among others, demanded acceptance of a mobile Earth.

Four conferences seem to have been of particular importance. The first held in 1926 by the American Association of Petroleum and published in 1928 condemned the ideas of poor Wegener (who was present) to oblivion for the next 30 years. The second, organised in Tasmania in 1956 by Carey and E. Irving, and published despite much opposition in 1958, was influential in opening minds, while that arranged at the Royal Society of London in 1964 began the assembly of our modern interpretation and the emergence of the theory of plate tectonics. The last conference, held in Berlin in February 1980, fittingly commemorated Wegener's centenary by setting out a comprehensive account of the new ideas.

It has been a pleasure to review these happy memories. My only regrets are that space and lapses of memory made mention of all who contributed impossible. As befits a mighty subject, it has needed many Lilliputians to assemble our fascinating picture of a mobile Earth.

26 November, 1981

PART TWO

Setting the Clock Back

3

Measuring geological time

L. R. WAGER AND STEPHEN MOORBATH

Knowledge of the Earth's history has been valuably increased during the past two decades by the development of new techniques for determining the absolute age of rocks. Study of regional age-patterns should throw light on the process of formation of the continents.

The aim of geology is to obtain an understanding of the structure and history of the Earth, and of those processes which have given the surface of our planet the form it has today. The early geologist obtained an historical perspective from the succession of the sedimentary strata and from the evolutionary sequence shown by the animal and plant fossils they contain. The fossils, when present, also allow correlation of the sedimentary strata across the whole world. A relative time scale based on this stratigraphical and palaeontological evidence was produced early in the last century and has been used with successive improvements ever since. However, there was no adequate way of deciding the absolute ages to be attached to the relative time scale. Neither was it possible to correlate vast thicknesses of unfossiliferous Precambrian rocks from one region to another.

Rutherford, as long ago as 1906, suggested a method for determining the absolute age of minerals by means of natural radioactivity. Lead was apparently being produced from uranium and thorium by radioactive decay at a characteristic rate which followed a well-defined law. Therefore the amount of the lead daughter product relative to the amounts of uranium and thorium parents could provide a measure of the length of time available for the decay of the parent element since its incorporation in the rock or one of the constituent minerals. Boltwood followed up Rutherford's suggestion; he analysed minerals for uranium, thorium and lead and obtained an estimate of their age, assuming that the sample was a closed system with regard to the parent and daughter isotopes – in other words, no uranium, thorium or lead had been added to or taken from the sample during its lifetime. He showed that the history of the Earth was to be measured in thousands of millions of years rather than in the tens or hundreds of millions of years as postulated previously from the rate of cooling of a once molten Earth to its present temperature, or from the rate of present day sedimentation in the sea compared with the observed thicknesses of the sedimentary rocks. On the basis of the uranium–thorium–lead method, the only one available to the early workers, an absolute time scale was constructed by Holmes and others which is surprisingly similar to that now accepted on the basis of recent investigations.

In 1938 a big step forward in age measurement became possible as a result of Nier's development of a mass spectrometer which measured quantitatively the ratios of the abundances of the individual isotopes which constitute a chemical element. It was known that the two isotopes of natural uranium, namely, uranium-235 and uranium-238, decay respectively into lead-207 and lead-206 and that the natural isotope thorium-232 decays into lead-208 (see Table 3.1), and although in each case the radioactive decay takes place in a complicated series of steps, the overall disintegration rate was well established for each series. Mass spectrometry provided accurate measurement of the amounts of all these isotopes, with the result that three independent ages could be obtained for a particular mineral, one from the ratio lead-206/uranium-238, another from lead-207/uranium-235 and another from lead-208/thorium-232. The ratio lead-207/lead-206 also gives an age, but it is not independent of the others. Thus for an age determination it is not sufficient only to analyse the mineral for uranium, thorium and total lead, but the lead must also be analysed isotopically. The mass spectrometry of lead is nowadays a relatively simple matter, although the apparatus is elaborate and costly. Many refined analytical techniques, both chemical and physical, are also available for the determination of uranium, thorium and lead, mainly as a result of post-war progress in the nuclear sciences. These can cope with amounts of these elements ranging from the per cent level to less than one part per million.

Most uranium and thorium minerals incorporate ordinary lead into themselves at the time of their formation, and this complicates the determination of the lead produced by radioactive decay.

Table 3.1 Naturally occurring radioactive isotopes used in age determinations

Parent isotope	Daughter isotope	Half-life* in millions of years	Useful geological age-range	Occurrence
Uranium-238	Lead-206	4510	Older than 10 million years	Uranium and thorium occur in local concentrations only. Minerals include uraninite, pitchblende, zircon, monazite, etc.
Uranium-235	Lead-207	713		
Thorium-232	Lead-208	13 900		
Potassium-40	Argon-40 / Calcium-40	11 850 / 1 470	Older than 100 000 years	Widespread in many rock-types. Biotite and muscovite micas are most useful.
Rubidium-87	Strontium-87	47 000	Older than 20 million years	Rubidium is a relatively rare element. Micas most useful.

* The half-life of a radioactive species is defined as the time in which half of the amount initially present will decay.

However, the lead-204 isotope can be used to estimate the amount of ordinary lead contaminating the mineral, since this isotope is not produced by radioactive decay in the mineral. "Common" lead contains the isotopes lead-204, lead-206, lead-207 and lead-208 and the ratios may be regionally variable. If a uranium or thorium mineral contains common lead it is necessary to analyse isotopically lead from a uranium-free mineral such as galena (lead sulphide) which is associated with the radioactive mineral, and to use the proportion of the lead-204 isotope to correct for the common lead which was incorporated into the mineral during crystallisation.

In favourable circumstances four ages can be obtained for a single uranium and thorium mineral, but, unfortunately, these frequently disagree amongst themselves. Such "discordant" ages are not due to analytical error, but to complicated physico-chemical processes which have acted on the mineral in the course of geological time, changing the parent to daughter ratios. Only the rare mineral uraninite (uranium dioxide) tends to give "concordant" ages, that is, agreement between the four ratios; when this is the case it must be the true age of crystallisation. Methods are also available for deciding which of the discordant ages are likely to be the nearest to the truth.

The uranium–thorium–lead method has the considerable disadvantage that minerals suitable for analysis are rare and mainly confined to localised zones of mineralisation. With the refinement of analytical techniques two new methods have come into use in recent years which can be used on commonly occurring rock-forming minerals. The first is the potassium–argon method depending on the radioactivity of potassium-40, which constitutes 0.0119 per cent of ordinary potassium. This undergoes a dual radioactive decay, in which 12.4 per cent decays to argon-40, the remainder to calcium-40. Only the decay to argon can be widely used for geological dating, since calcium-40 is the commonest isotope of normal calcium and most minerals contain so much normal calcium as totally to obscure the comparatively small radiogenic contribution.

The potassium–argon method requires a complicated high-vacuum apparatus, since the amount of argon gas to be measured frequently does not exceed one ten-thousandth of a cubic centimetre and may be considerably less than one-millionth. The potassium mineral is melted in vacuo at 1400–1500°C, releasing the argon and other gases. The argon is purified by removing all the chemically active and condensable gases and it is then measured in a calibrated capillary with the help of a travelling miscroscope. The amount of argon can also be obtained by the so-called isotope dilution technique; after fusion an accurately known quantity of argon-38 is added and the resulting mixture purified and then analysed on a gas mass spectrometer. Comparison of the peak heights of isotopes 38 and 40 gives the amount of argon-40 present in the mineral in terms of the amount of argon-38 added. The potassium content of the mineral also has to be accurately determined, but since the amount is usually between 1 and 10 per cent of the weight of the mineral, standard chemical techniques are adequate.

The potassium–argon method is applicable over a wide range of geological time; it has been used for the oldest known rocks, over three thousand million years in age, and also for young rocks formed less than half a million years ago. It is most useful for silicate minerals of the mica type (biotite, muscovite, lepidolite) which apparently retain 90–100 per cent of the radiogenic argon within their crystal lattice if the mineral has not been subjected to appreciable metamorphism after its formation. The more common, and often co-existing mineral, potassium-feldspar (potassium aluminium silicate), retains only a variable and unpredictable amount of argon, so that it is not generally used if mica is available, since it gives only a minimum value for the age.

The latest method in the repertoire of the geochronologist is the rubidium–strontium method, which depends on the decay of the isotope rubidium-87, comprising 27.8 per cent of natural rubidium, to strontium-87. Measurements were first made in 1953, but the rate of disintegration of rubidium-87 (see Table 3.1) was not established with reasonable certainty until 1959, owing to its weak radioactivity. Most potassium minerals contain some rubidium, since the two elements are geochemically closely allied. The micas and potassium feldspars normally contain between 0.01 and 0.1 per cent of rubidium and may often be used successfully for the rubidium–strontium method. A difficulty in the method is that common strontium, like common lead in the uranium–lead method, is sometimes present to such an extent that it obscures the radiogenic strontium-87.

The rubidium–strontium method also requires analytical techniques of extreme sensitivity and precision, involving the chemical solution of the mineral, the separation of rubidium and strontium by an ion-exchange technique and the subsequent measurement of both elements with the mass spectrometer to obtain the amounts of rubidium-87 and radiogenic strontium present. In favourable cases, one part per million of radiogenic strontium may be measured with a precision of 2–3 per cent. The useful age range of the method extends from the very oldest rocks down to about twenty million years ago.

It is evident that the best age determinations are those made by a variety of methods on several different minerals from the same rock. If they all agree within the experimental error (which should not normally exceed 2–3 per cent of the determined age) there can remain little doubt that the value represents the age of formation of the rock.

It should be mentioned that the more widely known carbon-14 method of age determination differs in principle from the above methods. Carbon-14 is produced by cosmic ray reactions in the upper atmosphere, after which it enters the atmosphere as carbon dioxide, whence it is incorporated in plant and animal tissue. It is radioactive and decays with a half-life of 5570 years. By measuring the present radioactivity of once organic materials such as fossil wood, cloth, etc., and making certain reasonable assumptions, such as the constancy of the amount of carbon-14 in the atmosphere, it is possible to determine the date of death of the original organism. This method extends back over only the last 70 000 years and is mainly used in archaeological problems and very recent geological ones, such as changes of sea level, the waning stages of the ice age and so on. The carbon-14 method differs from the long-term dating methods in that the actual radioactivity of the specimens is measured and not the amount of the daughter element.

Geochronological measurements have been made on rocks of the three principal kinds which make up the Earth's crust. Firstly, igneous rocks which have come from depth in a molten condition and either solidified as large, coarsely crystalline bodies before reaching the surface, subsequently revealed by erosion, or which have emerged at the surface as lava flows from volcanoes; secondly, sedimentary rocks, usually laid down in the sea and derived by erosion of pre-existing land masses; and thirdly, metamorphic rocks, originally igneous or sedimentary material subsequently reconstituted and chemically changed by heat and pressure during tectonic and igneous activity. The age of an igneous rock as determined by the above methods is the time of crystallisation of the minerals forming the rock. Sedimentary rocks, being derived from pre-existing materials, cannot be dated unless minerals which formed at the time of sedimentation can be extracted from them. Such a mineral is glauconite (a complex potassium magnesium iron aluminium silicate) which has been successfully dated by the potassium–argon and rubidium–strontium methods. Age determinations on minerals from metamorphic rocks, such as micas and feldspar, usually give the age of the metamorphism and in some cases, by a suitable choice of minerals and combination of age methods, it is possible to obtain information on two or more metamorphic events which may have affected the rock at widely separated intervals of time.

Most metamorphic rocks exposed at the present time represent the long-eroded remnants of mountain chains whose formation was due to deep-seated stresses in the Earth's crust. Age determinations up to the present suggest the possibility that there may be a cyclic periodicity in such events. Abundant mineral dates have been obtained from many parts of the world in the intervals 2700–2500, 2200–2050, 1850–1650, 1480–1300, 1100–930, 600–480, 360–280 and 120–10 million years ago, though by no means all of these are represented in any one region. Important and reliable ages measured on rocks from various continents are presented in Figure 3.1 which shows the time of major periods of

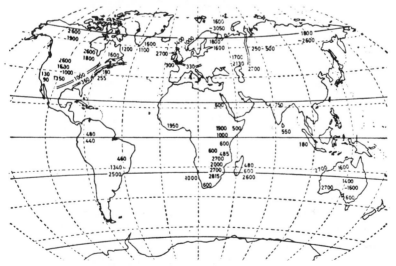

Figure 3.1 *Map of the Earth, showing principal age determinations. Some dated ancient mountain ranges shown by parallel lines*

Figure 3.2 *Recent age determinations of British rocks. Age in millions of years*

metamorphism, igneous intrusion or major ore deposit formation.

Until the last few years there have been few systematic age determinations of rocks in the British Isles. However, recent work in Oxford is providing interesting results. Very old rocks forming part of the Lewisian metamorphic complex of north-west Scotland are apparently nearly as old as any in the world, giving 2700 million years by the rubidium-strontium method (see Figure 3.2). Most of these rocks were affected again by a large-scale metamorphism about 1400 million years ago. Another complex mountain building process in the northern and central Highlands seems to have culminated about 420 million years ago, but dates have also been obtained from these rocks indicating that an earlier history will probably be decipherable. The ages of certain granites in the Southern Uplands, in the Lake District and in south-west England, the position of which are reasonably well fixed on the

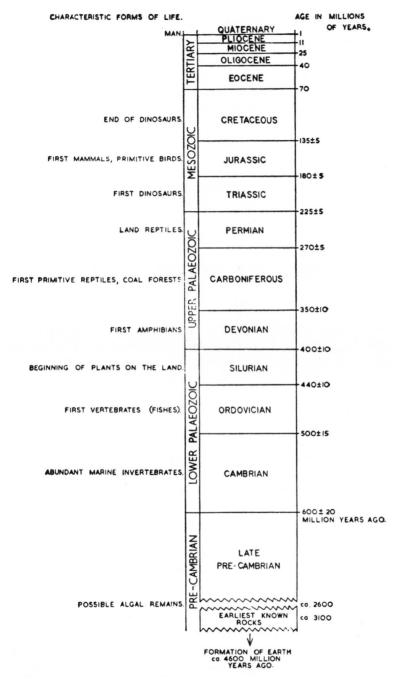

Figure 3.3 *Geological column with thicknesses proportional to time*

relative time scale, have also been investigated. The two northern granites, Shap and Cairnsmore of Fleet, were intruded about 390 million years ago and thus are a little later than the major Highland metamorphism. The sout-west England granite, the intrusion of which formed one of the last events of the great Hercynian mountain-building period, has an age of 290 million years and this date is confirmed by uranium minerals asociated with them.

The oldest known rocks of the Earth are just over 3000 million years old and the search continues for still older rocks. Meteorites have been shown by some of the above age methods to be about 4600 million years old. Since it is generally assumed that meteorites were formed at the same time as the Earth and the other planets, it seems that no rocks are yet known which were formed during the first third of the Earth's history.

Future study of regional age-patterns may be expected to give information on the actual process of formation of the continents, in particular whether they have been formed by successive accretion around original nuclei or whether they formed part of a single, large mass which disrupted and drifted apart. The latter hypothesis of continental drift is coming back into favour as a result of recent measurements on the direction of magnetism in rocks.

It seems possible that certain alga-like structures in ancient limestones found in the Zimbabwe–Tanzania shield area are truly organic. The rocks are some of the oldest known in Africa, their absolute age being about 2600 million years. If the alga-like structures are accepted as organic, then life has existed on the Earth for the astonishingly long period of 2600 million years. Really well-preserved and undoubted fossil animals are only known from Cambrian and later rocks, younger than about 600 million years. Postcambrian geological time, originally subdivided on the fossil evidence, can now be given fairly satisfactory absolute dating. The latest time scale, based on the 1959 paper of Professor Holmes, is reproduced in Figure 3.3. It is likely to be near the truth, having been compiled from measurements by all methods on materials with well-defined stratigraphical relationships.

The mass spectrometer and high-vacuum apparatus are instruments of a very different kind from the geologist's hammer, but they provide new means of tackling the old geological problem of the long and complex succession of events in the history of our planet.

14 July, 1960

4

Greenland yields the oldest rocks in the world

Preliminary exciting results from the Oxford Isotope Geology Laboratory indicate that some samples of granitic gneiss from east Greenland are the Earth's most ancient. A team working with Dr S. Moorbath have obtained a rubidium–strontium age of 3980 ± 170 million years (*Earth and Planetary Science Letters*, vol. 12, p. 245).

Previously, evidence for rocks older than about 3400–3500 My, has been sketchy, with an age for a Minnesota gneiss of 3550 My, while the few other reported great ages – between 3200 and 3500 – have not been those of primary crustal rocks. The Oxford workers believe that they have found the earliest evidence for a true granitic crust, implying that, by that time, the Earth's materials had begun the differentiation into granite magmas.

Dr Moorbath and his colleagues point out that their results now extend the terrestrial time-scale back beyond the apparent lunar melting event revealed by Apollo 12 and 14 rocks at 3600–3700 My; and the Apollo 12 ages of 3200–3300 My. Their samples, however, nowise resemble any lunar material so far. They maintain that if, as some planetary scientists have proposed, formation of the Earth's core released enough heat to melt the outer 1000 km of the mantle, such a holocaust must have occurred before 4000 My ago. The age of the Earth itself is placed at no more than some 4500 My.

"MONITOR", 2 December, 1971

5

World's oldest minerals found in Australia

Minerals in rocks from Western Australia are now being claimed as contenders for the title of the oldest known components of the Earth. Geologists in the School of Earth Sciences at the Australian National University have dated crystals of zircon at around 4200 million years old.

The team, led by Professor William Conpston, has measured the age of the zirconium silicate minerals in rocks from Mount Narrayer by radioisotope analyses. The technique employed was an advanced form of the one used in dating Moon rocks. It involved comparing the ratios of uranium to lead in the samples, as the former degrades into the latter.

Hitherto, the oldest known components of the Earth were those detected at Isua in West Greenland, in 1971 and dated at 3800 million years, by Dr Stephen Moorbath and a group at the University of Oxford. The Isua material is still the oldest known rock on Earth because the Oxford group dated it, and not just its component minerals. The Mount Narrayer rock as such is of a sedimentary nature and was deposited some 2800 million years ago. The fact that they contain older minerals indicates that a source region was being eroded as the rocks were being laid down. This would suggest there were solid crustal rocks on Earth 4200 million years ago.

Dr Moorbath told *New Scientist* that the news from Australia comes as a pleasant surprise. Some of the rocks brought back from the Moon are known to be 4500 million years old. It is probable that the Earth is just as old as the Moon.

"THIS WEEK", *19 May, 1983*

6

Fossil records of nuclear fission

R. L. FLEISCHER, P. B. PRICE AND R. M. WALKER

The authors' discovery that some common minerals preserve the traces of the natural fission of atomic nuclei occurring millions of years ago has led to the establishment of a new means of dating rocks and objects. Meteorites may preserve records of extremely rare cosmic-ray reactions.

If you dip a piece of ordinary mica in concentrated hydrofluoric acid for a few minutes, rinse it off and then examine it in a microscope at a magnification of a few hundred diameters, it is quite likely that you will see "fossil fission tracks" – tiny etched-out tubes about 0.001 cm long and even smaller in diameter. If you are fortunate in your selection of mica, you may see a grouping of tracks as photogenic as that in Figure 6.1. Many other minerals also contain fossil tracks which can be made visible by the proper choice of "developing agent". For example, Figures 6.2 and 6.3 illustrate the appearance of tracks in hornblende, a dark brown, weakly magnetic mineral, and in macusanite, a glass of volcanic origin found in Peru.

What are these little tracks, how were they discovered, and what kind of information can they give us about the minerals in which they are stored?

First of all, the tracks represent the paths in the mineral taken by fragments of uranium nuclei which have spontaneously undergone fission (each splitting into two heavy fragments) at some time during the Earth's history. In this atomic age it is hardly necessary to point out that the fission process involves the release of an enormous amount of energy, and when a uranium atom embedded in a mineral splits in two, the fission fragments lose their kinetic energy partly by ionising atoms in their path and partly by knocking atoms out of their normal positions. Each fragment thus

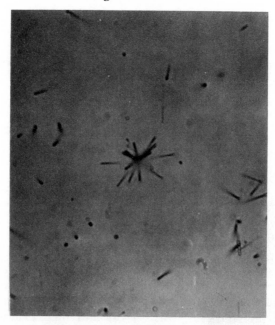

Figure 6.1 *Fossil tracks resulting from the spontaneous fission of uranium atoms in an ordinary piece of biotite mica × 950. The crystal was etched for 20 seconds in concentrated hydrofluoric acid. The star-shaped pattern of tracks radiates from a microscopic impurity containing a large number of uranium atoms*

ploughs out a trail of damage a few atoms in diameter and about 0.001 cm long. About four years ago, Dr E. C. H. Silk and Dr R. S. Barnes at Harwell bombarded a thin flake of mica with uranium fission fragments and, with the aid of an electron microscope, were able to see the submicroscope trails.

It then occurred to us that minerals which had not been deliberately irradiated might also contain trails of damaged material. As early as 1900, Lord Rutherford and others had been intrigued by the "pleochroic haloes" surrounding radioactive inclusions in mica. These haloes of discoloration are the accumulative effect of radiation damage by the alpha particles emitted from uranium and thorium in the inclusions. We therefore began our search for natural trails of fission events in mica, and a little over a year ago were successful in finding fossil tracks in the vicinity of a pleochroic halo (*Nature,* vol. 196, p. 732, 1962).

The observation was made possible by our discovery that fission-fragment trails can be etched out by certain chemical reagents, leaving fine hollow tubes or "tracks". If etching is continued the tracks grow until they can be seen in a low-power microscope. This makes it possible to find fission events from individual uranium atoms as easily as from the much larger uranium concentrations that are responsible for pleochroic haloes.

We have found that tracks in minerals can be created either when the nucleus of a uranium atom *spontaneously* fissions or

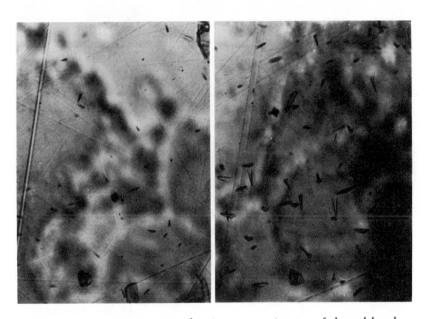

Figure 6.2 *Fission tracks in a specimen of hornblende mined from a quarry in Texas × 1100. To determine the age of a mineral such as this, two steps are required. (Left) A freshly prepared surface is etched and the number of fossil tracks in a known area counted in a microscope. (Right) The sample is mailed off to a nuclear reactor, exposed to a known number of neutrons, then re-etched and the number of new tracks in the same area is counted. The new tracks result from neutron-induced fission and tell us how much uranium is in the sample. From the ratio of fossil tracks to new tracks, it is possible to calculate that the oldest tracks in this hornblende were formed 1075 million years ago – a result in good agreement with the geological age of the quarry*

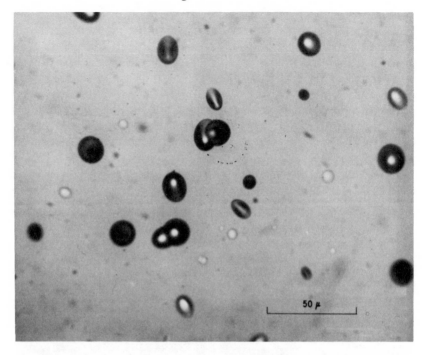

Figure 6.3 *Fossil tracks in an ancient glassy stone formed during the eruption of a Peruvian volcano. The event occurred 4.3 million years ago. The tracks, which were developed after a one-minute etch in hydrofluoric acid, are cone-shaped and have quite a different appearance from those in mica. × 500*

when a cosmic ray collides with the nucleus of a heavy atom such as lead and *induces* it to fission. We will discuss these origins in turn.

Spontaneous fission is very rare but occurs at a predictable rate given by the laws of radioactive decay. Let us imagine that we have a cube of mineral one cm on a side and containing one part per million of uranium by weight. We can expect that one atom of uranium will fission about every three years. About once every 1000 years a uranium atom will fission so close to the surface that the damage trail will intersect it and will be revealed by the etching treatment. If we know the uranium content of a mineral, we can then relate the number of fossil tracks intersecting a surface to the age of the mineral (as described in the caption to Figure 6.2).

Since nearly all minerals contain at least trace amounts of uranium, the discovery of fossil fission tracks appears at first sight to provide a simple and generally applicable method of mapping out the ages of the rocks on the Earth's crust. There are, however, a number of "ifs". Fossil track counts should provide the time since solidification of a rock *if* the temperature of the rock has been low enough over its entire lifetime to prevent atoms jumping around so much that they "erase" the tracks; *if* the number of uranium atoms in the rock has not changed over its entire lifetime; and *if* there are enough tracks in the sample to count in a reasonable time. The solid bar in Figure 6.4 shows the minimum ages of materials that we can study conveniently, depending on the

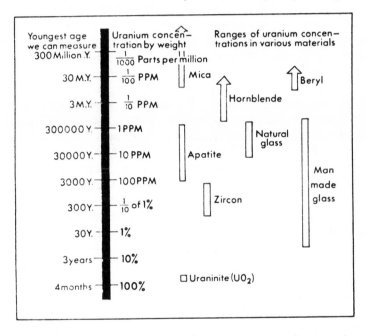

Figure 6.4 *The chart shows the approximate limits of our ability to measure ages as a function of uranium content. For example, if a piece of mica contained only 1/1000 part per million of uranium, then there would be too few tracks to count conveniently if it were less than about 300 million years old. Since the oldest glassware made by man dates back less than 5000 years, we see from the chart the only man-made glass objects we can hope to analyse contain more than 100 parts per million of uranium.*

uranium contents. If the material is rich in uranium, many tracks will be formed in a short time. Some of the materials with which we are working are shown, with vertical bars to indicate typical uranium contents.

In the last few years geochronologists have developed very powerful methods of dating certain minerals – those which contain large amounts of a radioactive isotope such as uranium-238, potassium-40, or rubidium-87. By analysing the products of radioactive decay with a mass spectrometer, they are able, in the most favourable cases, to date potassium-rich minerals as young as 500 000 years; and they can date rubidium-rich and uranium-rich minerals (zircons) which are older than 100 million years – or perhaps those which are as young as 10 million years if the mineral is a uranium ore. Looking at Figure 6.4, we see that a study of fossil tracks should enable us to date certain very young minerals in a region of the time scale which is inaccessible to the existing dating techniques.

In order to evaluate our fossil-track method, we have been making age determinations of a number of different minerals taken from geological regions which have been well dated by mass-spectrometric techniques. Some of our recent results are shown in Figure 6.5. The mica age determinations were made jointly with E. M. Symes of our laboratory, Professor D. S. Miller of Rensselaer Polytechnic Institute, and M. Maurette and P. Pellas of the University of Paris. Included in the graph are some measurements of several man-made glasses of known ages. These were studied in collaboration with Dr Robert Brill of the Corning Glass Museum. The youngest sample was a piece of mica placed next to a sheet of uranium for six months to simulate a uranium ore. About 400 fission fragments entered the mica from the uranium during this time, enough to make an accurate measurement of the "age" of the composite.

We are very optimistic about the accuracy and usefulness of fossil tracks as a chronometer for geologically "young" materials – those less than about 100 million years old. Minerals which are older than this have been giving mixed results. We tend to err on the young side when we date old micas, and the same is true of an old Canadian apatite crystal which we have analysed. On the other hand, our age of a Texas hornblende – 1075 million years – agrees well with the "known" age.

In order to learn more about the discrepancies in the upper right corner of Figure 6.5, we have heated several minerals and glasses

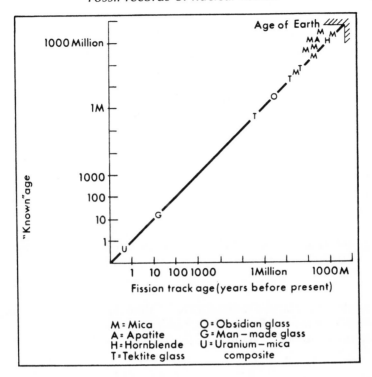

Figure 6.5 *Comparison of ages determined by fission-track analysis with ages measured by mass spectrometry, or otherwise known. The old micas which lie to the left of the line may have been heated or have picked up uranium since they were originally formed*

to various temperatures and have studied the disappearance of fossil tracks. We find that some minerals, such as quartz and hornblende, retain tracks at higher temperatures and for longer times than others, such as mica and apatite. Over long periods of time we expect minerals such as the former, which are most resistant to track erasure, to contain the most reliable information about their time of formation. This is suggested, but not proved, by the limited data in Figure 6.5. Minerals which are less resistant to track erasure, if properly calibrated, will reveal information about the time or severity of heating episodes during their history.

Let us now look at a specific application of fossil-track analysis to an old and controversial geophysical problem – the origin of tektites and impactites. Tektites are strange glassy objects that are found by the thousands in so-called "strewn fields" scattered across the Earth. Impactites are slag-like glassy rocks that are usually discovered near huge craters and are believed to have been formed from rock melted by meteoritic impact. Tektites are sufficiently rich in potassium to enable geochronologists to establish three age groups: 34 million years for the North American tektites, 15 million years for the Czechoslovakian tektites, and 700 000 years for Far Eastern tektites. Age determinations on most impactites have been impossible up until now.

We have determined fossil track ages for the three sets of tektites which confirm the ages obtained by potassium decay and, in addition, we have measured ages of impactites from as widely separated locations as Tasmania, the Libyan desert, and Quebec. The surprising result is that each of these impactite ages falls into one of the three tektite age groups. We conclude that tektites and impactites have a common origin; in view of their wide-spread spatial distribution and of the association of impactites with meteorite craters, we also conclude that some extra-terrestrial event was responsible for their creation.

Further information about their origin can be obtained by considering the second type of fossil tracks. If a material such as a tektite or a meteorite has been exposed to cosmic rays for a long time, it will contain, in addition to natural fission tracks, those induced by cosmic-ray interactions with the nuclei of heavy elements. When a tektite enters the Earth's atmosphere, the resultant heating may create a rim or flange from the flow of molten glass. In this material any tracks there originally are erased, whereas the inner core remains cool enough to retain any tracks that are present. We have compared track densities in the core and at the leading edges of several tektites and have seen no difference. This observation enables us to conclude that they cannot have travelled through outer space for more than a few thousand years.

Piecing all the evidence together, we see that three catastrophic events were responsible for the formation of the various tektites and impactites; that the objects are so widely scattered that they must have originated outside the Earth; and that the tektites had a short excursion through space before entering the Earth's atmosphere. An explanation which is consistent with this evidence is that assorted debris thrown off the Moon's surface by the explosive impact of giant meteors has found its way to Earth on at least three occasions and with sizes ranging from small droplets (tektites) to large chunks heavy enough to produce craters and form impactite glass.

The study of fossil particle tracks has really just begun. Only a few of the many varieties of terrestrial minerals have been examined. Meteorites – because thay have been exposed to cosmic rays for a long time – are obvious candidates for study. The first natural tracks in a meteoritic mineral – olivine – were recently found at the University of Paris in a joint experiment with M. Maurette and P. Pellas. In order to make quantitative studies of the production rates of cosmic-ray induced tracks, we are now bombarding various minerals with artificial cosmic rays in the Cosmotron particle accelerator at Brookhaven National Laboratory. An irradiation time of 30 minutes in the beam of three GeV protons simulates a cosmic-ray exposure of 100 000 years in outer space or of many millions of years on Earth.

Variations in cosmic-ray track distributions can, in principle, give us information about surface erosion rates of terrestrial rocks, of large meteorites, and eventually of the Moon's surface. Also, if we think of a meteoritic mineral as a kind of "cloud chamber" which has been exposed to cosmic radiation for a large part of the lifetime of the solar system, we see that we have a chance of finding a track of a very rare particle or nuclear event. To us this is an important incentive for intensive study.

In an age when both the equations and the equipment of science tend to become more and more complex, it is encouraging to note that it is still possible to discover a phenomenon like the existence of fossil tracks, which is basically simple, both in concept and in fact. It is perfectly straightforward for the amateur scientist, equipped with only a bottle of acid and a simple microscope, to reproduce our original discovery and to observe, in an ordinary rock, nuclear events that took place a thousand million years ago.

13 February, 1984

7

Fission track dating comes of age

LAURA GARWIN

Two physicists from the General Electric Research Laboratory in Schenectady, NY, proposed in 1963 a novel method for dating rocks, which involved the counting of radiation-damage tracks formed in the sample over long periods of time by the spontaneous fission of uranium-238. At the beginning of the month, and 21 years after P. B. Price and R. M. Walker's original paper, 60 geologists, physicists and statisticians from 15 countries met just across the Hudson River from Schenectady, in Troy, NY, for the 4th International Fission Track Dating Workshop. The results presented confirm that fission track dating (FTD) is a welcome addition to the stable of radiometric dating techniques available to the geologist. Also clear is the utility of fission track studies in determining the temperature that rocks have experienced over long periods of time.

The "age" of a mineral, as determined by any radiometric dating method, is not necessarily the time elapsed since the mineral first crystallised. Rather, each technique dates the time at which the relevant "geological clock" was reset to zero; that is, the time at which the daughter product (in the case of FTD, the fission tracks) began to accumulate. Fission tracks in minerals consist of defects in the crystal lattice caused by the passage of the energetic, charged fission fragments; subsequent heating causes the defects to heal. If a mineral is subjected to a sufficiently high temperature for a sufficiently long time, any existing fission tracks will disappear, and the FTD "clock" will be reset. This is the key to the investigation of thermal histories by FTD. The temperature required to anneal tracks varies with the mineral being dated, and it is customary to quote for each mineral a "closure temperature", below which tracks will be stable over geologically relevant spans of time (1–1000 million years). For example, the closure tempera-

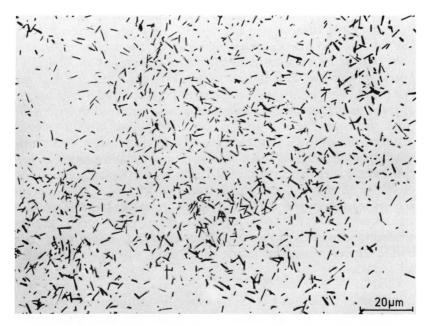

Figure 7.1 *Fission tracks in muscovite mica, revealed by etching in hydrofluoric acid*

tures for apatite, zircon and sphene (the three minerals most commonly used for FTD) are approximately 100°, 225°, and 300°C.

Perhaps the most exciting results presented at the workshop related to the low-temperature end of this spectrum. Andrew Gleadow and his colleagues at the University of Melbourne observed that the temperature interval over which track annealing occurs in apatite (70–125°C) is almost identical to that required for the maximum generation of liquid hydrocarbons from organic matter (a process known as "maturation"). The fission tracks in apatites separated from a possible source rock thus reveal whether that rock has been subjected to sufficient heating to generate oil. Other techniques exist which can answer this question, but the advantage of the fission track method lies in the fact that one can obtain not only the maximum temperature reached, but also the time at which the heating occurred.

Gleadow's group has studied the lengths of fission tracks in apatites from deep wells in sedimentary basins in Australia. The annealing of a track proceeds gradually along its length, so that a track shortens progressively at a rate dependent on temperature until it finally disappears. The apatite ages and the mean track lengths showed the expected decrease as the well temperature increased with depth, until all tracks disappeared at about 125°C. Furthermore, the distribution of track lengths about the mean broadened with depth so that each sample displayed a distribution of track lengths which was characteristic of its position within the track annealing zone. When these characteristically broadened distributions of track length are found in rocks which are currently at temperatures below those required for track annealing, one can conclude that the rocks concerned have spent a significant amount of time in the track annealing zone, and therefore in the temperature range required for hydrocarbon maturation. That this technique is of more than academic interest has been demonstrated by the birth of Geotrack International, a company owned by the University of Melbourne which does fission track work for oil companies.

Charles Naeser of the US Geological Survey presented two examples of the use of FTD in mineral exploration. He emphasised that, at least in the southwestern United States, all the mineral deposits which have some obvious visible expression at the surface have been found and exploited, so that it has now become necessary to look for more subtle surface expressions of mineralisation, at depth. That is, we need to look for evidence of temperatures having been raised by mineralising fluids at some depth below the rocks that are now exposed at the surface – temperature increases which in most cases have long since decayed away. Naeser also reported that the spatial distribution of partially and totally reset zircon and apatite fission track ages has revealed the location of deepseated mineral deposits in two mining districts in Colorado. Once again the advantage of FTD over other geothermometers is revealed in its ability to date the thermal anomaly, and thus the mineralising event.

Another major application of FTD is in the study of the uplift of mountain chains. Due to the increase of temperature with depth in the Earth, a rock which is raised to form part of a mountain cools through the various mineral closure temperatures, recording the times at which each temperature was reached. Fission track ages of samples from different altitudes can be used to derive the rate at which uplift occurred. An entire morning's session was devoted to this topic, and the contributions spanned five continents. Randall Parrish, of the Geological Survey of Canada, emphasised the important assumption which often goes unquestioned (or even unstated) in deriving a rate of uplift from fission track or other radiometric ages; namely, that the geothermal gradient has remained constant during the time interval spanned by the mineral ages. The slow rate of diffusion of heat in the Earth makes this assumption unlikely to hold for uplift rates of more than about 0.5 millimetres/year, a rate which is often surpassed in regions of active mountain-building. Parrish stressed that FTD must be integrated with thermal modelling and with any constraints provided by the regional geology, in order to construct a reliable history of the thermal and tectonic evolution of an area.

The workshop discussed in detail the fundamental phenomena underlying the FTD technique. For instance, in order to quote a closure temperature for a given mineral, it is necessary to understand the kinetics of track annealing in that mineral. Our understanding of the processes of creation, destruction and revelation of fission tracks in natural materials continues to develop, but the workshop participants demonstrated that it is possible to produce reliable ages for rocks from a wide variety of geological settings, and to derive thermal information which could not have been obtained by other radiometric dating techniques. It is clear that on its 21st birthday (and like the oil it is helping to find), fission track dating has reached maturity.

"MONITOR", 23 August, 1984

8

Geology and the computer

DANIEL F. MERRIAM

Geologists, habitually unaccustomed to quantitative thinking, are passing from an era characterised by data collection to one in which objective syntheses of their accumulated information is becoming possible. This gentle revolution is yet another facet of the computer age.

Geologists, long adjusted to thinking in terms of millions of years in regard to age of the Earth, can now make effective use of computers that perform operations in millionths of a second in geological problem solving. The advent of the computer is revolutionising geology, just as it has other scientific disciplines. Application of the computer to geology, however, has been slow because of the reluctance of geologists – mostly due to their traditionally qualitative training – to accept more rigorous, quantitative methods. Thus the complete potential of the computer has yet to be realised in the Earth sciences, and it may be many years before a full evaluation of its impact can be made.

It is clear at this stage, however, that computers will relieve the geologist of many routine and menial tasks, and allow him more time to think about problems and means of their solution. In addition, the computer permits analytical methods to be utilised, which would have been too tedious or even impossible to apply by the older methods. The use of the computer in treating geological data, therefore, can be grouped into two general categories: *data processing* and *problem solving*.

Since the late 18th-century days of James Hutton, a Scotsman and founder of modern geology, geologists have accumulated data on all aspects of the Earth. At first, storage and assimilation of these data was relatively simple; but as more and more information has become available, maintenance and utilisation of the data have become increasingly difficult. Now, however, concomitant with arrival of the computer, geology has arrived at a stage where synthesis is possible.

Data storage and retrieval have been readily adapted by the petroleum industry where literally thousands upon thousands of borehole records are available. Each record can contain as many as a million bits of information on location, geology, engineering, production and hydrology. In the United States, several independent and cooperative ventures for converting basic well information into machinable form are beginning their operation or are contemplated. Such well data systems permit rapid access to and recovery of tremendous amounts of information in any desired sequence or form. Data systems of a less pretentious nature can likewise be devised for more particular purposes. Various small systems for storage and retrieval of geochemical, palaeontological, and stratigraphical data are now in operation.

It is impractical, if not impossible, for geologists to evaluate the flood of present-day publications, especially when they are increasing at an increased rate. To help alleviate this dilemma, Keyword-In-Context (KWIC) indexes – that is indexes in which keywords in the title are alphabetically arranged in different combinations – are being prepared for different publication and reference lists to facilitate library work. KWIC is designed for a fast visual scan to convey the contents of the document being indexed and consists of the index proper, bibliography, and author index.

For many years, some branches of geology, such as sedimentology, have made use of descriptive statistical methods. Only very recently, however, have probability, regression, and correlation techniques been used on a large scale. There are three reasons why geologists in the past did not generally use these more advanced methods: they were not quantitatively inclined; the numerical data necessary may not have been available generally; and the techniques for practical utilisation of the methods did not arrive until the advent of the computer. A serious sample-distribution problem is unavoidable because geologists by necessity are restricted to sampling at sites where the rocks are present and available. This limitation results in biased sampling and presents unique problems in determining any statistical inference from non-random samples.

It should be emphasised here that quantitative methods are of particular significance because, with the same set of data, different investigators can arrive at the same conclusions. The main advantages of quantification then, as noted by R. R. Sokal and P. H. A. Sneath in their book *Principles of Numerical Taxonomy* (1963, W. H. Freeman and Co.), are repeatability and objectivity. Also, it is not humanly possible to consider more than a few variables at a time; geology, however is characterised by many variables and the number of variables that can be considered by making use of large computers is many.

Some of the principal multivariate methods used in the Earth sciences include factor analysis, time-series analysis, and regression or trend analysis. Let us have a look then at these methods.

Factor analysis, developed early in this century by psychologists, is a method for finding common underlying factors in many interdependent variables. Its application to geological problems has been almost entirely confined to determining the interrelations of chemical, lithological, and biological constituents in recent and ancient sediments, thereby grouping sampled localities into similar areas; and in this way meaningful patterns are revealed. These patterns in turn are used to interpret the environmental history of the sedimentary unit in question.

Such a study has been made of a thin (up to five feet thick), persistent, marine limestone – the Americus Limestone – in order to interpret the environment in which it was laid down by reconstructing the distribution of sediments on the sea floor in Early Permian time (about 250 million years ago). Rock specimens were sampled along an approximately 250-mile outcrop in eastern Kansas and Oklahoma. Quantitative information on their constituents was obtained by studying thin sections of the rocks and by chemical and spectrographic analyses. The raw data were first transformed to give each constituent approximately equal weight when comparing localities, then appropriate coefficients were computed for subsequent factor analysis. The factors extracted from the Americus Limestone data are shown in Figure 8.1a. Major changes are indicated (labelled "phases"), which are related to distribution of constituents determined from the original data. With this information, the depositional environment of the unit was re-created. It is shown in Figure 8.1b.

Time-series analysis permits analysis of information collected at successive intervals of time. Most time-series applications in geology do not involve time diretly but instead, are dependent on a relative timescale because the actual (or absolute) time is not known. Time-series analysis has been applied to geological observations taken along a traverse or stratigraphic section.

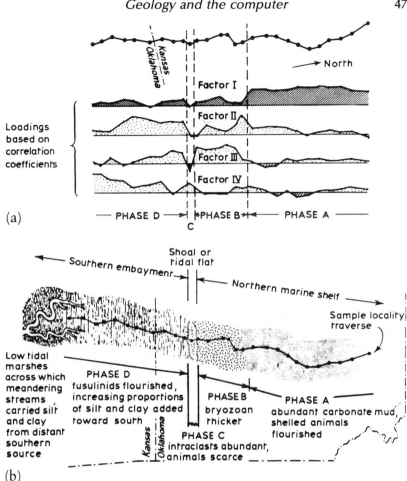

Figure 8.1 (a) *Graph of "loadings". The changes reflect variations in all constituents used in calculating coefficients and mark the boundaries between divisions labelled phases.* (b) *Map showing interpretation of "phases" of the Americus Limestone*

Observations may be measurements concerning the porosity of the rocks, their chemical or mineralogical composition, grain size, fossil distribution, rock fabric, bed thickness, or other aspects. The object of time-series analysis is to introduce order into the irregular or erratic appearance of the data thus allowing meaningful interpretation (Figure 8.2). By analysing a time series it is possible to separate it into basic components: secular or long

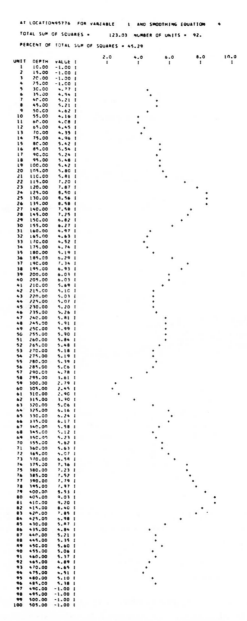

Figure 8.2 *In this "smoothed" time-series curve, the time (along the horizontal axis) needed to drill successive five-foot intervals of rock (shown vertically) is displayed. Harder units, such as limestone, take longer to drill and the corresponding values are deflected to the right*

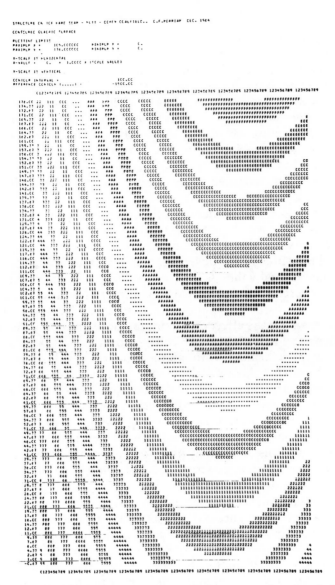

Figure 8.3 *Computer contoured map of second-degree trend surface filled to configuration of the Hard Seam coal in part of the Nottingham-Derbyshire Coalfield. The area extends northward from Nottingham to just north of Worksop and includes approximately 650 square miles. Top of map is north; edges of alternate bands of letters and numbers with blanks represent contour lines; contour interval = 100 feet*

range, cyclic; and random or irregular variations. Depending on the problem, one or other of the system components can be emphasised by near elimination or minimisation of the others.

Harmonic analysis can be applied to the cyclic component of a time series to determine its periodicity. This technique has been successfully used in studying varved sediments – those which show a seasonal layering in response to climatic factors such as temperature and precipitation. The sunspot cycle of 10 to 13 years, in addition to other cycles of about 60, 85, 170, and 180 years, has been recorded. The longer cycles reflect climatic changes.

Trend surfaces are mathematically computed plane or gently curving surfaces which represent general, regular trends. The difference between the computed value of the trend surface at a point and the value of the observed, actual surface at that point is termed the residual value. If the trend surface is thought of as the regional or large-scale component, then the residual value can be considered the local or small-scale component. Removal of the regional trend has the effect of amplifying or re-enforcing the local component and making it more conspicuous.

Many papers dealing with application of trend surfaces to geological and geochemical data have been published; most of these papers are concerned with the distribution of various constituents in igneous and sedimentary rocks. One interesting application has been the simulation of geological structure – the shape and slope of the rock layers – with trend surfaces. Quite frequently minerals, including petroleum, are associated with a particular type of geologic structure, and delineation of this kind can aid in mineral exploration.

Trend-surface analysis can be applied to regional, areal, or local structural features (Figure 8.3). Preliminary results indicate that application to structural entities or parts of them, especially basin-

Figure 8.4 *Mathematically computed "trend surfaces" show the overall or regional trends of geological structures. The first-degree surface is the plane which is the closest fit to the actual structure. Second-degree trend surfaces are bowl- or saddle-shaped. Later trend surfaces become increasingly complex. The so-called residual values are derived from the difference between computed and actual values at any given point and represent the local component of the geological structure.*

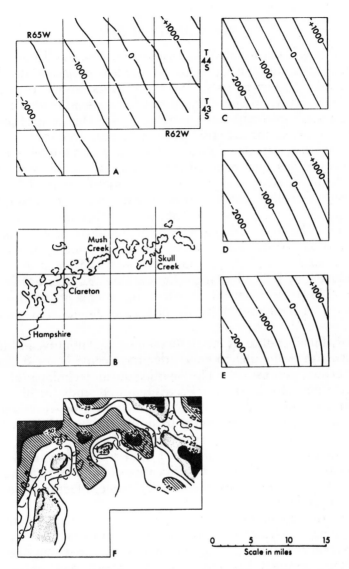

Above is shown an example of how trend-surface residuals can be used to accentuate the relationships of geological structure to oil and gas fields in part of north-east Wyoming: **A** *actual structure—the lines are underground contours giving the shape of a layer of rock;* **B** *map of oil and gas fields;* **C** *first-degree trend surface;* **D** *second-degree;* **E** *third-degree;* **F** *relation of first-degree residuals to oil and gas fields* (D. E. Merriam and J. W. Harbaugh, 1963, *Oil and Gas Journal*, vol. 61, no. 47).

like or down-warped areas is most satisfactory. Results of a trend-surface analysis can be used to predict projected depths to particular geological rock layers and units within an area; to delineate structural discontinuities; and to extend better "geological guesses" into adjacent unknown areas. This type of analysis does not reveal features that can not be perceived in the original data by close scrutiny. What it does do is accentuate structural features of less than regional magnitude and, in this way, brings out details of structures previously unnoticed (Figure 8.4).

Four variables can be treated with trend "hypersurfaces", imaginary surfaces that are "above" or "beyond" an ordinary surface. The four variables can be represented by 1, a variable, such as oil gravity, water salinity, ore grade, etc., 2, stratigraphic depth, and 3 and 4, geographic position. Such a study was conducted by J. W. Harbaugh (*Kansas Geological Survey Bulletin*, 171, 1964) on crude-oil gravities in an area in south-eastern Kansas, where he found that the oil gravity did not conform to the configuration of the rock strata.

An almost endless list of applications of quantitative techniques utilising the computer to good advantage could be cited, and the list grows continually. The mathematical technique of auto-correlation can be used effectively in a study of the cyclic alternation of rock sequences, and that of cross-correlation to correlate one vertical section with another. So-called discriminant functions can be used to place individual factors into one or other of a number of groups, which have been defined previously on the basis of various attributes. Classifications of different kinds can be erected using coefficients of similarity between factors as the data for cluster analysis.

Pattern recognition and geological modelling or simulation are becoming important in studying natural processes. Methods of map comparison, long difficult or impossible, are being studied by reducing complex situations to simple patterns and analysing the components. Decision-making theory is being used in the petroleum industry to help make decisions where many variables are involved.

To help gather and disseminate computer programmes of interest to workers in the Earth sciences, an international clearing house to be known as GEOCOMP has been proposed by J. W. Harbaugh (*Geotimes*, vol. 8, no. 7). If this proposal becomes a reality, then one of the difficulties encountered in working with computers – lack of usable programs – would be lessened. Another problem, cost of computer operation, has been substantially reduced in recent years with the many commercial and academic installations available at a nominal cost. Many exciting and interesting discoveries and developments can be predicted for the future in geology.

20 May, 1965

9

The origin of the Solar System

B. J. LEVIN

The author surveys recent theories: most of them agree that the planets were formed cold from a cloud of dust and gas. There is still, however, disagreement about the details – in particular about the Moon's origin and the nature of the Earth's core.

About 30 years ago the theory of the origin of the Solar System put forward by Jeans became invalid under the critical onslaught of Russell, Pariisky and Spitzer. Subsequently, scientific thought turned away from the idea that planets may have had their origin in the hot agglomerations of solar gases, and began to favour the idea that the planets were accretions of cold matter. The classical Kant-Laplace theory of the origin of planets from matter scattered over the entire region of the Solar System was re-examined in the light of contemporary knowledge of physics.

In 1943 Weizsäcker, the German physicist, and O. J. Schmidt, the Soviet mathematician and polar explorer, put forward their hypotheses: Weizsäcker's starting point being a Sun surrounded by a gas and dust cloud, while Schmidt presupposed the capture, by the Sun, of a swarm of solid bodies and particles from interstellar clouds. Both scientists presumed the gradual growth of planets by the agglomeration of the available matter.

In order to explain the existence of a regularity in planetary distances (see Table 9.1), of which Bode's law is a rather unsuccessful expression, Weizsäcker assumed the existence of fairly regular systems of giant vortices in the protoplanetary cloud. This artificial assumption was probably the chief shortcoming of Weizsäcker's hypothesis and prevented its acceptance.

Schmidt explained the regularity of interplanetary distances quite plausibly as a result of competition for matter between neighbouring planetary nuclei. Two such nuclei, if their orbits

Table 9.1 The Planets

Bode's Law states that the mean distances of the planets from the Sun are in the ratios 4:4+3:4+6:4+12:4+24, etc. As the table shows, it is only approximately correct.

Group	Planet	Mean distance from Sun (Earth=1)		Mass (Earth=1)	Mean density (Earth=1)
		Measured	Predicted by Bode's Law		
Terrestrial Planets	Mercury	0.39	0.4	0.057	~1
	Venus	0.72	0.7	0.82	0.92
	Earth	1	1	1	1
	Mars	1.52	1.6	0.107	0.71
Asteroids	—	Various c. 2.8	2.8	Small	—
Major Planets	Jupiter	5.20	5.2	318.4	0.24
	Saturn	9.54	10.0	95.3	0.13
	Uranus	19.19	19.6	14.6	0.25
	Neptune	30.07	38.8	17.3	0.4
—	Pluto	39.52	77.2	?	?

were particularly close to one another, would quickly empty the space between them; then the planetary nucleus which was nearer to the Sun would have continued to grow, mainly as a result of adding to itself bodies arriving from nearer the Sun, while the outer nucleus would add to itself bodies arriving from farther away. Each captured body would add to the planet not only its mass, but also its angular momentum. Consequently, the angular momentum of the first planetary nucleus would decrease, and that of the second increase, so that their orbits would move away from each other.

The simultaneous application to the process of accretion of the laws of conservation of angular momentum and of energy, while taking into account also the heat dissipated in inelastic collisions, enabled Schmidt to account for the planet's tendency to rotate on its axis.

Schmidt concluded that the Earth was originally cold and that its interior became hot gradually by accumulating heat from radioactive decay. It was a crucial result for geology and geophysics.

In 1950 Professor L. Gurevich and Professor A. Lebedinsky published their researches, which were very important for the further elaboration of Schmidt's ideas. They pointed out that the difference in the mass and composition of the inner terrestrial planets and the outer major planets could be explained on the assumption that the solid bodies from which they accumulated were not captured by the Sun but were formed in its vicinity from the gas and dust of the protoplanetary cloud. In the inner parts of the cloud, which are warmed by the Sun, small dense planets formed, consisting of non-volatile matter; while in the outer, very cold parts of the cloud huge planets of low density were formed, consisting mainly of volatile matter, which is far more abundant in space than non-volatile matter.

Following the publication of this work, Schmidt began to assume that the Sun captured a gas-dust cloud, rather than a swarm of meteorites. However, as he himself remarked, the fundamental features of the process of planet formation from this cloud would remain the same, even if it had been formed simultaneously with the Sun, as many astronomers suppose.

As a result of their friction with the gas, the dust particles in the protoplanetary cloud were bound to have small random velocities, and therefore would collect in the central (equatorial) plane of the cloud, forming there a disc of dust. The dust then agglomerated into many bodies, of the size of the present-day asteroids, possibly as a result of dust particles sticking together, but more likely as a result of local gravitational instability. At first these asteroidal bodies rotated round the Sun in circular orbits in the plane of the disc. However, as they became sufficiently large to exert gravitational effects on each other, their motion gradually grew less regular, and their orbits became elliptical and inclined. Collisions ensued which sometimes resulted in the combination and sometimes in the fragmentation of the bodies. Eventually the asteroidal bodies and their fragments formed the small number of planets which exist at present. Asteroidal bodies continue to exist to the present day only in the wide space between the orbits of Mars and Jupiter.

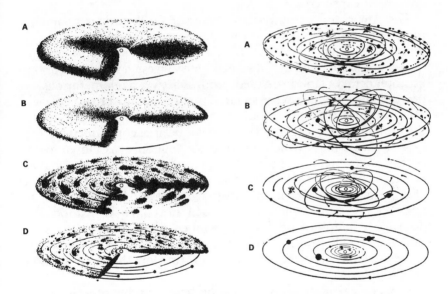

Figure 9.1 *(Left) The formation of asteroidal bodies from a gas-and-dust cloud in the vicinity of the Sun. A and B, dust collecting into a disc; C and D, disintegration of the disc into agglomerations which turn into asteroidal bodies. (Right) The formation of planets from a swarm of asteroidal bodies. A, increase in eccentricity and inclination of orbits as a result of mutual gravitational perturbations; B and C, accretion of asteroidal bodies and their fragments to form planets; D, the planetary system today*

There is one important unsolved problem in this picture of planet formation: the giant planets, particularly Jupiter and Saturn, contain much hydrogen, which freezes at a temperature of 4–6° K. It is not clear whether the temperature of the outer parts of the protoplanetary cloud could have been quite low enough to allow the accretion of hydrogen.

The dust disc must have been extraordinarily opaque, but solar rays reached it laterally, scattered by the gas component of the cloud which was not very flattened. The answer depends on the unknown configuration of the shadow which the inner parts of the disc cast on its outer parts. If it were shown that the prevailing temperature conditions could not permit the freezing of hydrogen, the growth of the giant planets would have to be accounted for mainly by a process of accretion whereby gas as well as dust would have been captured.

The American astrophysicist Professor G. P. Kuiper imagined the formation of planets from the protoplanetary cloud in a different manner. Like Weizäcker, he wanted first of all to explain the regularity of planetary distances. He suggested in 1949 that the protoplanetary cloud divided into massive agglomerations – protoplanets – which were so large that they almost touched each other in their rotation round the Sun. Their transformation into the contemporary planets would have been accompanied by the loss of excess mass which, in Kuiper's opinion, was blown away by the solar "wind". At first Kuiper supposed that the smaller bodies in the solar system – satellites, asteroids and comets – had also been formed from the protoagglomerations. However, as agglomerations of small mass could not have withstood the destructive action of tidal forces, Kuiper, during later researches, saw that small bodies, and perhaps even the small terrestrial planets, would necessarily have been formed by the accumulation of solid particles.

Kuiper's calculations on which he based the hypothesis of the formation of massive protoplanets were criticised by E. Ruskol at a symposium on the origin of the Earth and the planets which took place in Moscow in 1958 during the general assembly of the International Astronomical Union. Kuiper agreed with this criticism, but still retained his opinion that giant planets had been formed not by growth, but by getting rid of excess mass.

At the same symposium Professor F. Hoyle put forward his new hypothesis on how the protoplanetary cloud was created in the course of the formation of the Sun from a contracting nebula.

Separation of matter, as a result of very fast rotation, soon ceased as a result of magnetic coupling between the separated matter and the Sun. The rotation of the Sun would then be retarded, and the separated matter would move away from it and spread over the space now occupied by the planets. This theory, published in 1960, is the first concrete expression of the idea of the common formation of the Sun and the protoplanetary cloud. It contains a few points of uncertainty, but on the whole it looks very promising.

The idea of the importance of electromagnetic forces in the formation of the planets had been expressed even earlier by Professor H. Alfvén, the Swedish originator of the science of magnetohydrodynamics. According to his hypothesis these forces always played a decisive part. However his treatment of the problem required a number of rather artificial conditions, and even then much remained unexplained. According to Alfvén, atoms in the gas cloud which was falling towards the Sun were gradually ionised, and then their fall was stopped by the solar magnetic field; of the more abundant elements the first and most distant element to become ionised and retarded should have been iron, and the last and nearest, hydrogen. However, the chemical composition of the planets shows exactly the opposite.

Nevertheless, the consideration of electromagnetic forces, as Hoyle showed, could be decisive for solving the problem of the distribution of the angular momentum between the Sun and the planets.

Professor H. Urey, who won the Nobel Prize for his heavy water research, produced important results when he attacked the problem of the origin of the planets in 1950. Unlike most cosmogonists, who based their theories mainly on the peculiarities of planetary motion, Urey, as a physical chemist, paid more attention to the analysis of the chemical composition of the bodies in the Solar System, particularly of the Earth and of meteorites, for which more data were available.

On the Earth, and even more so in meteorites, there is a considerable deficiency of volatile elements, including the heavy ones such as xenon and krypton, while they are abundant in the Sun, and in space in general. At the same time, these bodies contain not only non-volatile elements, but also elements of medium volatility, such as zinc, cadmium and arsenic, in "normal" cosmic abundances. Urey showed that here was convincing proof that the Earth and the asteroids, of which meteorites are fragments, had been formed, not as a result of the dispersion of massive gas or gas-and-dust condensations, but by the accumulation of solid particles which had not been subjected to heating. Therefore Urey, who at first tried to explain the regularity of planetary distances using Kuiper's protoplanets, began later to object firmly to the idea, at least applied to the terrestrial planets. His later picture of the formation of planets is very similar to that described above in connection with the work of Schmidt.

However, there is one important difference between the views of Urey and those of the author of this article who, together with a group of colleagues, continues to develop Schmidt's theory after the latter's death. This difference arises from a different approach to the nature of the Earth's dense core: Urey considers that it is mainly iron, while I, in accordance with Ramsey's theory, think of it as consisting of silicates transformed into a dense metallic state under high pressure. If the Earth's core is ferrous, the difference in the mean density of terrestrial planets and the Moon indicates a difference in their metallic iron content. For ten years Urey has been looking in vain for a process which could explain this difference. He has changed his viewpoint several times. However, if the Earth's core consists of metallised silicates, Venus, Earth, Moon and Mars have almost the same composition, and only Mercury consists of denser matter, which is naturally to be explained by its proximity to the sun.

These two points of view lead to a divergence of our ideas on the origin of our Moon. According to Urey, the lesser density of the Moon shows that, compared with the Earth, it has hardly any metallic iron. Therefore his opinion is that it was formed somewhere far from the Earth, in different conditions, and was subsequently captured by the Earth. But the circular shape and the small inclination of the lunar orbit are arguments against its capture, which would be a rather improbable event anyway. However, if, starting from Ramsey's hypothesis, the comparatively great density of the Earth is explained by high pressures inside it, while assuming the composition of the Earth and the Moon to be essentially the same, it becomes natural to presuppose their simultaneous formation from the same substance. My colleague, Ruskol, showed that during the agglomeration of the Earth, a swarm of particles of sufficient mass would necessarily be formed around it as a result of inelastic collisions of particles in its vicinity, and the Moon could form later by accretion of these particles. This process of Moon formation seems simple and natural.

So far, however, neither theoretical calculations for magnesium oxide nor experiments on compression by shock wave of dunite, a mineral containing much olivine and considered an adequate representative of the internal matter of the Earth, have shown any sudden increase in density under pressure of 1½ million atmospheres, which correspond to pressures at the boundary of the Earth's core. In 1960 B. Mason, a geochemist at the National Museum in New York, expressed the opinion that the best sample of protoplanetary matter was a rare type of meteorite – a carbonaceous chondrite containing many minerals of the chlorite–serpentine type. This opens new hopes of verifying Ramsey's hypothesis.

Carbonaceous meteorites are additionally interesting because they contain not only amorphous carbon, but also various complex hydrocarbons. Having investigated some similarities between these meteorite hydrocarbons and those of butter and organic sediments, the American scientists Nagy, Meinstein and Hennessy recently came to the sensational but ill-founded conclusion that they had obtained proof of the existence of life outside Earth. Actually, hydrocarbons in meteorites can be of entirely inorganic origin, and I am convinced that this is the proper explanation. Therefore, if carbonaceous meteorites are samples of protoplanetary substance, complex hydrocarbons existed on Earth from the very beginning and could have been the basic materials for the origin of oil and life.

But there is another possible way in which hydrocarbons could have arrived on Earth. Comet spectra show that cometary nuclei, which are icy bodies, consist, *inter alia,* of some so far unknown hydrocarbons.

Even now comets are bound to collide with the Earth now and then, and it is possible that the fall of the Tungusca "meteorite" in Siberia in 1908 is an example of such a case. In the dim past there have been incomparably more numerous comets which entered the region of the Earth's orbit. During the final stage of the formation of giant planets, when there were still many icy bodies around them, the latter acquired very extended orbits due to gravitational perturbations from Jupiter, Saturn, Uranus and Neptune, and they frequently entered the region of the terrestrial planets. If our Solar System had been visited at that time by an astronaut from a distant planetary system, he would have witnessed a remarkable spectacle – several bright comets could be seen in the sky almost always, some of them having two or even three tails.

This was also the time when Professor J. H. Oort's distant "cloud of comets" was formed. Comets which went away from the Sun to 50 000–100 000 astronomical units became affected by stellar perturbation, changed their orbits, and no longer approached the Sun and its planets. As they were not affected by the heat of the Sun they became preserved (as it were) in a cosmic refrigerator from which, now and again, these same stellar perturbations occasionally extract them.

The quoted examples show that the problem of the origin of the Solar System closely involves questions of celestial mechanics astrophysics, geophysics and geochemistry, and even organic chemistry and biology. In order to ascertain the actual process of the formation of the Earth and other bodies of the Solar System a joint analysis of all aspects of the problem is necessary. This is, of course, a difficult task, but results achieved during the past score of years show convincingly that the research is in the right direction.

8 February, 1962

10

The primitive Earth

PRESTON CLOUD JR

There is a good reason to suppose that the Earth's atmosphere and hydrosphere arose as the result of secondary processes occurring after the formation of the Solar System. If the Moon was captured by the Earth, the accompanying catastrophic events may have caused the initial outgassing responsible; in the subsequent evolution of the environment both living organisms and iron ores probably played leading roles.

It is the prevalent view nowadays that the Earth and Solar System originated by the gravitational contraction of a cold cloud of dust and gas. A close study of the ensuing evolution of the Earth's atmosphere and oceans reveals, however, that neither can be directly attributed to this primary process. To read this ancient and extended span of Earth history the geologist must make use of several highly diverse lines of evidence, and consider the interactions between living organisms, and geochemical and geophysical systems. When he tackles the problem in this way a coherent picture of events on the primitive Earth begins to emerge.

One of the most surprising chemical features of the terrestrial atmosphere is its great depletion in the noble gases as contrasted with their cosmic abundances. This fact alone seems to require that our atmosphere be of secondary origin. Either the Earth originated without an atmosphere, or it lost such an atmosphere in a later thermal episode. Nevertheless, the continuous existence of sedimentary rocks dating back for more than 3000 million years (three aeons), demonstrates the continuity of our atmosphere and hydrosphere over a period of at least that length.

At the same time, the existence of detrital grains of readily oxidised minerals in stream deposits only two aeons old means that the atmosphere recently contained no free oxygen. And the puzzle is further complicated by quantities of primary ferric oxide in marine deposits of the same age, which tells us that there *was* a source of oxygen within the large water bodies from which these deposits precipitated. Ferric-oxide-coated detrital sediments ("red beds") of non-marine origin as old as 1.8 aeons, moreover, indicate that by then free oxygen was accumulating in the atmosphere; and the subsequent evolution of life and sediments hints at later steps in atmospheric evolution. From such fragile threads we can weave a strong, if coarse-textured, tapestry of events on the primitive Earth.

How and when the atmosphere began

In fact, Earth scientists proposed an internal source for the terrestrial atmosphere long before anyone recognised the depletion of the noble gases. And they suggested that it arose from outgassing and weathering following the accumulation of the primordial rocks. But the compelling force of the evidence was not widely appreciated until 1951, when the American geologist W. W. Rubey published an incisive assessment of possible sources for the atmosphere and hydrosphere. Now the concept of accumulation of both from juvenile sources is generally accepted, and discussion focuses on the composition and time of origin of the primitive atmosphere and the amounts and changes in it of oxygen, nitrogen, carbon dioxide, and hydrogen – especially free oxygen, for which no primary source is available.

As to the time of origin of the atmosphere, an outside limit is the age of the Earth. Radiometric data show that the materials of which the Earth is composed originated about 4.55 aeons ago, at the time of origin of the Solar System as a whole, but they do not tell us exactly when these materials aggregated to form the Earth. A minimal age for that event is, of course, given by the most ancient terrestrial minerals dated. This age appears to be around 3.5 to 3.6 aeons. That, at least, is the age of the oldest radiogenic dates obtained on four intensively studied continents, and it seems that some very significant history event must have taken place at about this time. It also highlights the yet unsolved problem of what was happening for the first billion years or so of Earth history, assuming that accumulation of the planetary materials was completed not long after the big homogenisation event 4.55 aeons ago.

The point of interest here, however, is that, inasmuch as we do not have rocks older than about 3.6 aeons, there are few constraints on conjecture about the nature and origin of an atmosphere or hydrosphere that might have antedated that time.

Some constraints on conjecture

The oldest rocks that tell us something definitive about the early hydrosphere and atmosphere are sedimentary rocks in southern Africa believed to be somewhat more than 3.2 aeons old. They could not have originated in the absence of atmospheric weathering and a substantial hydrosphere. They also limit conjecture about the nature of the atmosphere beneath which they accumulated. In addition, the easily oxidised minerals uraninite and pyrite of detrital origin in various younger rocks combine with other evidence to indicate that the atmosphere between about 3.2 and 1.8 aeons ago could have contained little or no free oxygen. Other geochemical evidence implies that, contrary to the popular methane–ammonia model of the primitive atmosphere, there could have been little ammonia or methane in the atmosphere from about 3.2 aeons onward. The early atmospheric gases instead would have been those that are trapped in igneous rocks, or are juvenile components of volcanic and hot-spring gases – H_2O, CO_2, CO, N_2, HCl, H_2S, and a few other trace gases.

In order for an atmosphere and hydrosphere to begin, the outer Earth would have to undergo melting sufficient to release its "volatiles". Such melting might be produced by the combined effect of radiogenic heating, tidal friction, and the conversion of gravitational energy to heat. Or it might be the product of lunar capture, abetted by these same factors.

If, among the older geologic records, some provided clues to tidal amplitudes, we could test whether or not it was likely that the Moon was in orbit at any particular time – say as far back as 3.6 aeons ago. Such geologic evidence does exist, although not older than about 3.2 aeons. It includes dome-shaped sedimentary structures of algal origin known as stromatolites, plus other sediments indicating intertidal deposition.

Domal stromatolites that rise conspicuously above the surface on which they grow are found only between tides, where, among recent forms, they reach a maximum known relief of 0.7 metre. The West Australian geologist Brian Logan found that height (or

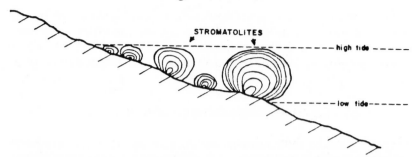

Figure 10.1 *Schematic profile showing relation of domal stromatolites to intertidal zone*

amplitude of stromatolite layering) was determined by tidal range and position in the intertidal zone as in Figure 10.1. Fossil stromatolites can similarly give a clue to tidal ranges in the geological past.

Now it happens that stromatolites of pre-Palaeozoic age often have much greater amplitudes than younger ones (2.5 to 6 times as great), and over large areas.

The implication one may draw from this and other data is that tidal amplitudes too great to be accounted for by the solar component alone existed at least two aeons ago; that tides generally were probably greater then than now; and that the Moon was, therefore, already in orbit and closer to the Earth between 2.0 and 0.6 aeons ago than it is now.

The evidence of the more than 3.2-aeon-old Swaziland System of south-eastern Africa is also consistent with the Moon being already in orbit around the Earth that long ago. Rocks at the top of this system have the characteristics of extensive intertidal deposits, implying tidal amplitudes too great for the solar component alone.

If the Moon was in orbit 3.2 or more aeons ago, and if it did not originate at the same time as the Earth, perhaps its origin had someting to do with the 3.5 to 3.6-aeon thermal event implied by the oldest known metamorphic and granitic rocks and by lead-isotope data (Figure 10.2).

If the Moon was captured, tidal friction sufficient to induce subcrustal melting would have been likely. Melting, in turn, would have promoted outgassing and accretion of atmosphere and hydrosphere, together with a general resetting of geologic clocks. Any pre-existing terrestrial atmosphere and hydrosphere would have been lost at that time, and a new or first atmosphere and hydrosphere started. It is also conceivable that the postulated lunar capture and accompanying thermal episode could have given rise to a temporary lunar atmosphere and hydrosphere.

Chemistry of rocks and fossils

Thus our atmosphere and hydrosphere may have evolved, with additions, from one that began with pervasive melting about 3.5 to 3.6 aeons ago (Figure 10.2). To account for microfossils more than 3.2 aeons old, life must have originated soon after. Such life would have been anaerobic in the absence of oxygen, and dependent on external food sources. However, it could not have continued, giving rise to the observed evolutionary record, without the appearance of an organism that could manufacture its own substance – an autotroph, probably a photosynthetic autotroph.

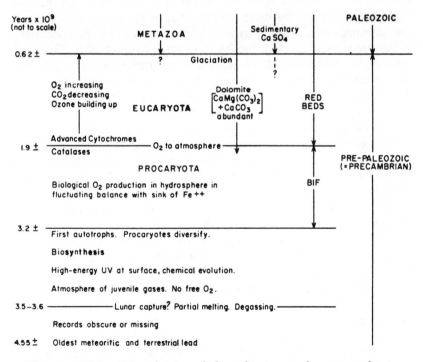

Figure 10.2 *Biospheric, lithospheric and atmospheric evolution on the primitive Earth*

And photosynthesis is the most likely process by which free oxygen might be generated in quantities sufficient to produce an oxygenous atmosphere.

In the absence of suitable oxygen-mediating enzymes, however, free oxygen is lethal to all forms of life! When oxygen-releasing photosynthesisers arose, therefore, they would have faced the problem of disposing of oxygen in such a way as not to burn themselves up. Unless oxygen-mediating enzymes preceded the origin of oxygen-releasing photosynthesisers, such organisms would have been dependent on an associated oxygen acceptor in the physical environment. This is where the ancient haematitic iron ores and red beds enter the story.

Ferrous ions whose oxidation produced the world-wide haematitic banded iron formation (BIF) between about 3.2 and 1.8 aeons ago may have been the oxygen acceptor. Nothing like the BIF – a feature of the old rocks on every continent – of any thickness or regional extent is found in younger rocks. The geochemical problem posed by the BIF is how to transport the iron in solution, under oxidising conditions, or to precipitate it under anoxidising conditions. This problem is resolved by the concept of a balanced relationship between organisms and BIF. The iron would be transported in solution in the ferrous state and precipitated as ferric iron on combining with biological oxygen. The banding suggests a fluctuating balance between oxygen-producing biotas and supply of ferrous ion.

The Earth's first biologically generated oxygen thus was locked in chemical sediments, and free oxygen did not appear in the atmosphere except in small quantities from photolytic dissociation of H_2O and CO_2 rapidly scavenged by the reduced substances then abundant in the atmosphere and at the surface of the Earth.

When efficient oxygen-mediating enzymes did arise, this balance would have collapsed. Primitive algae equipped with such enzymes could spread as widely through the hydrosphere as light penetration and ultraviolet-shielding mechanisms would permit. They would have swept the ocean free of ferrous ions, and oxygen would accumulate in excess and begin to escape to the atmosphere. The last great episode of BIF, about 1.8 to 2.0 aeons ago, may mark such an event.

What would have happened when oxygen began to build up in the atmosphere? At that time, in the absence of an ozone screen, solar high-energy ultraviolet light would have been able to reach the surface of the Earth. Some of the molecular oxygen (O_2) would have been converted to atomic oxygen (O) and ozone (O_3). Iron would have been retained in the weathering profile of the Earth in the ferric state. Because of the great chemical activity of O and O_3 even a low rate of transfer of O_2 to the atmosphere would cause extensive oxidation of surface materials. Red beds should appear in abundance in the geological column at that time. The oldest thick and extensive red beds now known are about 1.8 to 2.0 aeons old – a little younger than, or overlapping slightly with, the youngest BIF. This date may mark the time in atmospheric evolution when free oxygen began to accumulate.

The appearance of atmospheric oxygen would also set the stage for the emergence of a new type of organism. Palaeontological evidence implies that until then all organisms consisted of procaryotic cells, those lacking a nuclear wall and being incapable of mitotic cell division and genetic exchange as a normal accompaniment to reproduction. The presence of free oxygen, even in small quantities, was presumably followed by the evolution of the eucaryotic cell, with nuclear wall, well-defined chromosomes, mitotic cell division, and the capacity for sexual reproduction and genetic recombination as the usual mode of replication.

How fast did oxygen accumulate in the atmosphere once it started, and what were its biological consequences? At first the green plant photosynthesisers would still be confined to protected sites in sedimentary mats, or where they would not be circulated into surface waters, until such time as an efficient ozone screen built up to exclude DNA-inactivating radiation in the neighbourhood of 2600 angstroms. The late Lloyd Berkner and his colleague L. C. Marshall found that this happens at about one per cent present atmospheric level (PAL) of oxygen. Both they and (earlier) the Canadian biologist J. R. Nursall, moreover, suggested that the appearance of differentiated multicellular animal life (Metazoa) was a consequence of the achievement of atmospheric oxygen concentrations sufficient to support a metazoan level of oxidative metabolism.

The appearance of the Metazoa in the geologic record, of course, does require two necessary, if not sufficient, preconditions. One is the origin of the eucaryotic cell, of which all metazoans are constituted. The other is a sufficient level of free oxygen – although perhaps closer to three than to one per cent PAL.

Now the oldest rocks in which eucaryotic fossils are known are about 1.2 to 1.4 aeons old, although eucaryotes may have made their debut before this. The precondition of the eucaryotic cell, therefore, was satisfied well before the dawn of the Palaeozoic about 620 million years ago; and I have elsewhere documented the conclusion that there are as yet no records of unequivocal Metazoa in rocks of undoubted pre-Palaeozoic age. This suggest that sufficient free oxygen may have been triggering the event.

The apparent abruptness of early Metazoan evolution may be partially explained by a polyphyletic origin – a wave of multi-cellular forms derived from different possible metazoan ancestors almost simultaneously. Since, moreover, all ecologic niches that could ever be occupied by Metazoa were then unoccupied, adaptive radiation probably contributed to diversification of the metazoan root stocks. This biological revolution may have taken place over an interval of say 100 million years – roughly equivalent to the time indicated for chemical evolution leading to the origin of life itself, and somewhat more than that required for the Cenozoic diversifications of the mammals following extinction of the dinosaurs.

What evidence other than the geologically rapid evolution of the earliest Metazoa at this time might suggest that the dawn of the Palaeozoic approximately coincided with the appearance of a level of free oxygen adequate for metazoan metabolism? When ozone reached a level sufficient to exclude the DNA-inactivating ultra-violet radiation at about one per cent PAL of oxygen, it would open up the surface waters of the entire hydrosphere to occupation by photosynthesising phytoplankton. That, then, could generate a large increase in the amount of oxygen in the atmosphere. A sudden big increase in atmospheric oxygen might correlate with several features in lithospheric evolution. The increase in oxygen would presumably have been paralleled by a decrease in the carbon-dioxide blanket serving to reflect heat radiation. The result could have been a temperature decrease sufficient (other conditions being suitable) to account for the widespread late pre-Palaeozoic glacial deposits recognised by many geologists. Such an oxygen increase would also be consistent with the observed abrupt increase in abundance of sedimentary calcium sulphate in basal Palaeozoic rocks and an episode of oxidative enrichment of the BIF in late pre-Palaeozoic or earliest Palaeozoic time.

14 August, 1969

11

Evolution in environments

ROLAND GOLDRING

Concern is mounting over man's increasing influence upon his environment, and its possible repercussions. This article illustrates how other living creatures have had major effects on their surroundings. It forms a natural sequel to Professor Preston Cloud's article "The primitive Earth".

The extent to which man is rapidly changing the natural environments on the surface of the Earth is all too familiar today. Over the regions he has actively colonised by farming, by the destruction of forests, by chemical pollution of the atmosphere and so on, he has greatly modified relatively stable environments that existed formerly. Man's actions over the past few hundred years mark the beginning of a major revolution in the pattern and types of environment present on the Earth's surface. As yet, this revolution is most clearly in evidence in the land; oceanic environments, and those of the larger seas, are so far scarcely affected.

When one looks at the geological record of the past 1000 million years it is interesting to see that this not the first time that organisms have affected the Earth's surface on so vast a scale. Indeed, it may be argued that, in the past, natural environments have been modified to an even greater degree by bursts of organic activity, though never at quite the pace of the present revolution.

Land plants began to evolve slowly about 400 million years ago, but during the Carboniferous period (some 300 to 350 million years ago) there was a prolific increase in their abundance and diversity, which led to the formation of major coal deposits. Professor Fairbridge, of Columbia University, has argued that the result was an immense removal of carbon dioxide from the atmosphere. The increase in oceanic alkalinity that followed possibly contributed to the unusually great extinctions about 225 million years ago, which were most marked among marine invertebrates.

The effects of organic growth and activity are recorded in many of the sediments that were deposited in the formerly existing environments. Such sediments can be traced back beyond 3000 million years to about the time when life itself began to exercise a slowly increasing control over the natural environments.

Sedimentary rocks cover much of the Earth's surface and it is from these rocks that fossil fuels stem, and in which many natural resources occur – such as iron ore and basic building materials like gravel, clay, and limestone for bricks and cement. An appreciation of the effect of organisms in past and present environments is important for geologists' attempts to interpret fossil sediments in terms of natural environments and in prospecting for these resources. For example, although the oil and gas fields of Alberta are associated with 360-million-year-old Devonian "coral reefs", the overall form of these reefs and the structure of the fossil elements is quite diferent from any present-day coral reef, calling for an evolutionary approach on the part of the oil-field geologist.

The two major controls which geologists generally hold to be responsible for the type and association of fossil sediments are climate, and the structural, tectonic, state of the source and depositional areas. One might contrast the deposits in deep ocean troughs opposite lofty, tropical mountain ranges with those in shallow warm seas opposite dry deserts. Climate and structure are subject to the basic physical constants and though they might combine to give many types of environments, there is as yet no evidence that the constants have changed over the past 3000 million years, though the gravitational force at the surface of the Earth must have decreased, assuming the Earth to have expanded. Thus hot arid deserts must always have looked the same.

Organic control must undoubtedly be recognised as a third major control and, moreover, one which is evolving with geological time. The effects of organisms on sedimentary environments may be grouped under four headings:

First, there is the vast bulk of sediment actively contributed by organisms in the form of macroscopic and microscopic shell and other skeletal materials (Figures 11.1 and 11.2). Sometimes this sediment will accumulate where the animals actually lived and died as, for instance, the rocks formed by the extensive calcite skeletons of an extinct group of sea-lilies (crinoids) during the Devonian and Carboniferous periods (Figure 11.2). In a bore into

Figure 11.1 *Block of limestone (full-scale) with skeletal remains of fossil Archaeocyathids; Lower Cambrian, Ajax Mine, Flinders Ranges, South Australia (inset, reconstruction of group)*

the Carboniferous Limestone at Filton near Bristol a few years ago almost 30 per cent of the 480 feet cored was of crinoidal limestone, the welded, fossilised skeletal debris of the crinoids. The Chalk of southern England and elsewhere in north-west Europe is often 99 per cent calcium carbonate – of organic origin. In the Isle of Wight, the Chalk is 1800 feet thick. The equivalent amount of sediment in Ordovician times, long before the evolution of the minute organisms that make up much of the Chalk, would have been only a few feet of claystone.

Second, organisms disturb the normal distribution of the sediments by diverting and retaining material. Nowhere is this more evident than in the Middle East where many of the oil fields are associatd with the reefs and lagoonal sediments which flourished during the Mesozoic and Tertiary periods. Without the

Figure 11.2   *Polished block of limestone (full-scale) from Carboniferous Limestone of Derbyshire, showing stem ossicles of fossil sea-lilies, crinoids (inset, reconstruction of complete camerate crinoid)*

organically constructed reefs there would be no lagoons. In an area where the Earth's surface is subsiding the lagoon is a vast sediment trap and a highly favoured niche for organic growth. There were no organically formed lagoons in Precambrian and early Palaeozoic times (Figure 11.3). Only relatively ephemeral sand bars thrown up by waves served to break wave attack. For the past 50 million years reefs have been very efficient structures leading to the formation of vast lagoonal areas and also protecting the shores.

In Britain we are familiar with tree-shaded river banks and the main function of the trees in protecting the bank against erosion seems somewhat incidental. The evolution of the prop-rooted, salt-tolerant, mangroves some 65 million years ago probably changed the appearance of all younger tropical deltas and many tropical coastlines. Around our own shores in modern times the effects of the hybrid grass, *Spartina townsendi*, since it appeared some 50 years ago, are obvious in fixing mud, leading to the silting up of many estuaries and creeks.

Third, living organisms inhibit erosion. They achieve this end in several ways, though perhaps the effect of plants slowing down rain run-off is the most important. Though not so strong in the early evolutionary stages of plants during the Devonian and

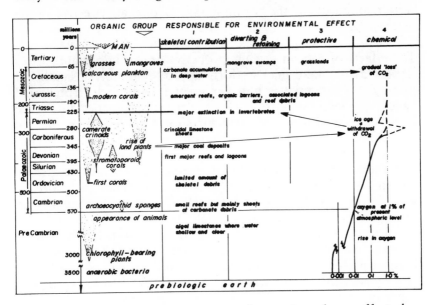

Figure 11.3 *How some groups of organisms have affected and modified environments through geological time. The model of oxygen evolution (right) is taken from Berkner and Marshall* (New Scientist, vol. 28, p. 418). *The oxygen level attained by the beginning of the Cambrian is now (1984) generally held to be about 10 per cent of present atmospheric level (rather than the 1 per cent shown in the figure)*

Carboniferous periods when most of the plants were growing in depositional areas (swamps, etc.) it has been especially significant since the beginning of the Tertiary period (Figure 11.3) when grasses colonised much of the land surface. We know the effect of removing the grass cover; the terrible look of the overgrazed badlands of North America and Australia; even the gullying which may begin on patches of cleared waste ground in temperate northern Europe.

Around the periphery of the oceans the sediment is mostly in a relatively unstable position with the surface being aggraded at one time of the year and degraded at another. Various organisms, particularly algea, polychaete works and mussels, are able to create sediment baffles, leading to deposition and stabilisation of sediment which would otherwise be hydrodynamically mobile. Some of the rock-boring organisms, conversely (particularly certain molluscs and the boring sponge *Clione*) are effective agents of erosion.

Fourth, organisms influence the chemistry of the environment. The effects of certain organic processes are so important in this respect as to make man's efforts appear miserably insignificant (though his potential is now far greater with nuclear power). I have mentioned the possible effect of plant upsurgence during the Carboniferous. More lowly structures form a second example; the planktonic, calcite-secreting organisms, including certain foraminifera, and the minute coccolith algae. These showed a tremendous burst in abundance during the Cretaceous period (about 100 million years ago). Previously, carbonate in the oceans was almost entirely deposited in shallower water, particularly on the continental shelves, by animals and plants favouring such conditions. Over geological time the continental shelves have been frequently uplifted, warped and, in places, folded. When uplift has been sufficient to expose the sediments to the forces of weathering and erosion the carbonates have been returned to the oceans in solution and recycled. However, the trend now is for calcium carbonate to be deposited on the relatively stable floors of the oceans.

Does this mean that the potential of the seas for the production of carbonate rocks is being gradually reduced because of permanent deposition in deeper regions of the ocean? If so this tendency might eventually lead to a decrease in carbon dioxide in the atmosphere and reduction in plant growth – in spite of the vast amount of carbon dioxide now being added to the atmosphere by the burning of fossil fuels.

I have discussed several of the major organic events in the Earth's history. Others are shown in Figure 11.3 and Professor Cloud has discused those in the Precambrian. Organic evolution has not been as orderly as might be generally thought and there are many perplexing instances where Darwinian factors do not seem to have been dominant. No good explanation is yet forthcoming for many of the major extinctions of animal groups. Often there is no obvious, more efficient group which replaced the declining group; in other instances no group seems to have taken immediate advantage of an apparently available ecological niche. For instance, at the base of the Cambrian period (about 600 million years ago), a group of organisms, known as the Archaeocyathida (Figure 11.1), with a stout calcareous skeleton and most closely resembling the sponges, evolved very rapidly and their fossils are now classified into eleven sub-orders (according to *The Fossil Record*, published by the Geological Society of London and the Palaeontological Association). Their remains account for hundreds of feet of limestone in Siberia and Australia. They had become extinct by the beginning the Middle Cambrian and the ecological niche they occupied in the warm shallow seas was not as effectively inhabited again for about another 80 million years when the earliest corals established themselves. During that period the environments to which they belonged presumably reverted almost to those existing in Precambrian times.

Man's actions are really only another stage in the evolution of environments – though, since he now has some measure of control, how he plays it is rather important for his own future. This discussion could now lead into science fiction but one might reasonably ask the question: From what it is feasible to discover about the evolution of environments over the past 1000 million years or so, what possible events might be forecast which would significantly affect man or lead to new patterns of natural environments?

Plants, or plants in association with groups of invertebrates, offer both the greatest potential and the greatest potential danger. In a relatively small way it has been possible to record the changing environments in Lake Kariba (see *New Scientist*, vol. 36, p. 750) where, for a time, mats of water fern were a definite danger to the control of the lake. Such effects are minute compared with, say, the extensive colonisation of the seas and even oceans by long-stalked sea-weeds with sea-anchor-like roots.

A rather more subtle danger is the plague of starfishes now preying on living coral in Australia and in the Red Sea. Starfish "control" of coral growth could lead to a reduction in the protective function of coral reefs and the destruction of coral islands. However, the lessons for geologists are those that can be learned from the present day and, as the editors of a recent geological text-book remarked, an appreciation of the nature of biological processes is essential in the training of a geologist. Without this, the history of the Earth's cover is meaningless, because environments are essentially the products of biological activity.

16 October, 1969

12

The evolution of the Earth's atmosphere

ANN HENDERSON-SELLERS

Interest in the evolution of the Earth's atmosphere is at a peak as a result of predictions that man's activities may be becoming a significant force in modifying the "natural" state of affairs. An international conference on planetary atmospheres is being held this week in Nice; here, one of the participants at that conference sets the scene by discussing how the Earth got the atmosphere it has today.

The fascinating topic of the evolutionary processes which produced the atmosphere we have around the Earth today is no longer a specialist preserve of the planetary scientists, but has become a puzzle of broader, and more practical, importance. The study of climatic change, both natural and man-made, is now at the forefront of the Earth sciences as many experts wonder whether conditions as pleasant for life as those of the past few decades are likely to persist. And a full understanding, and prediction, of climate depends on a full understanding of the atmosphere, where it came from, and how it got into its present state.

The interdependent cycles affecting life in the atmosphere and oceans (and in the longer term, influences from continental changes) combine in our planetary ecosystem, which seems to have serenely withstood 4.5 billion years (4.5 By) of traumatic upheavals. These include an increase in solar luminosity, volcanic activity, the formation of oceans, mountain building and continental drift, and the origin of life on the planet. It is difficult to obtain evidence relating to the prevailing climatic conditions over time scales comparable with the age of the Solar System itself; but something can be said about the value of the average global surface temperature and its changes over most of this time.

The present day temperature is 288 K (15°C), and is clearly

compatible with a hospitable environment for life as we know it. Geological evidence suggests that this global mean surface temperature has not varied outside the range 280–300 K throughout the Earth's history. Liquid water first appeared on our planet around 3.8 By ago, so this particular ecological niche seems to have been relatively stable for a very long time. Researchers' attempts to model mathematically the dynamic interactions between the Earth and its atmosphere have, one has to admit, been grossly inadequate. Nevertheless, two views seem to be emerging: one, that our present position on Earth is a happy, but statistically unlikely, chance situation; the second, a considerably less anthropocentric view, is that the planetary environmental evolution is inherently stable.

Experts now generally agree that the Earth and other planets near the Sun (the terrestrial planets – Mercury, Venus and Mars) lost any primordial atmosphere through solar heating early in their lives. The present atmospheres are thought to be a result of loss of gas from the mantle, including both volcanic outgassing and the vaporisation produced by the impact of meteorites. But the experts are not agreed on just what kind of atmosphere this outgassing produced in the first place.

The same degassing processes today would liberate mainly water vapour and carbon dioxide, so the simplest guess is that the Earth's atmosphere started out from a mixture of these two gases. But instead the usual argument is that the atmosphere produced initially was rich in gases like methane and ammonia – reducing gases similar to those dominating the atmosphere of Jupiter and the other giant planets today. This conventional view seems to be inherited from the early theories of the origin of life, which were based on the build-up of complex organic molecules from a reducing atmosphere and water, stimulated by sunlight and lightning discharges. Now, though, experiments have shown that complex amino acids can build up in an atmosphere dominated by carbon dioxide, and there is an extreme view (put forward by Sir Fred Hoyle and Chandra Wickramasinghe) that life might originate in interstellar space, not even requiring a planetary base (*New Scientist*, vol. 76, p. 402). In addition, both Venus and Mars are now known to have carbon dioxide atmospheres. The simplest explanation is that all three planets started out in the same way, with predominately carbon dioxide atmospheres produced by degassing. So how did the three planets – especially the Earth – get to their present-day states?

Table 12.1

	Venus	Earth	Mars
T_e(4·5 By ago)	360K	300K	245K

The most important parameter, by a long way, is the mean global surface temperature at the time when an atmosphere began to form. This determines where the water goes to, and that in turn determines the evolution of the planetary system from then on. Without an atmosphere, the temperature is simply the effective temperature, T_e, which results from the balance between incoming solar radiation and the rate at which heat is lost to space by the planet. The value appropriate for rocky planets in the orbits of Earth, Venus and Mars are given in Table 12.1 – these are the surface temperatures the three planets must have had 4.5 By ago.

Crucially, the temperature on Venus then was high enough for water to be kept in its vapour state; from the very beginning, the water vapour in the atmosphere of Venus must have trapped infrared radiation, eventually producing a runaway greenhouse effect which has made Venus a hot desert today. At the other extreme, on Mars, the water couldn't even melt, let alone evaporate, so the atmosphere remained thin, with water trapped in frozen reservoirs below the surface. On Earth, however, things were – and remain – more interesting.

The intermediate position of our planet resulted in temperatures which ensured condensation of water vapour released into the atmosphere, forming large ocean areas, permitting carbon dioxide solution and leading to the formation of sedimentary rocks as rain carried eroded material into the seas. With carbon dioxide removed from the atmosphere (see Table 12.2) the path of the future surface temperature evolution was determined.

Table 12.2 Inventory of carbon near the Earth's surface (normalised)

Biosphere marine	1
nonmarine	1
Atmosphere (in CO_2)	70
Ocean (in dissolved CO_2)	4000
Fossil fuels	800
Shales	8000 000
Carbonate rocks	2 000 000

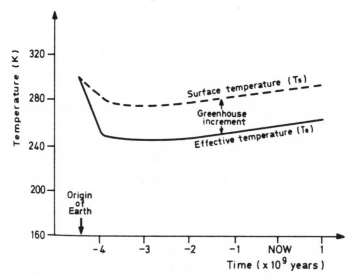

Figure 12.1   *Stable evolutionary track of the Earth's surface temperature. The "greenhouse effect" increases rapidly during the early stages of the evolution. T_s remains above 273 K throughout*

Once the atmosphere is established, the planet becomes more reflective to solar radiation, and also a better radiator of heat; the effective temperature drops, but later must increase again as the Sun's luminosity rises. Now, however, the average global surface temperature, T_s, depends not only upon T_e but also upon the greenhouse effect – the extent to which the atmosphere acts as a blanket, trapping heat that would otherwise be radiated into space. This depends on the kind of gases in the atmosphere, and how much there is of each (Figure 12.1).

Water is critically important for many biological, chemical and physical processes. It also controls temperature on Earth both by the greenhouse effect and by the extent to which snow and ice cover, or clouds, change the reflectivity of the Earth (its albedo). These feedback processes have ensured temperatures on Earth close to 290 K, permitting liquid water to dominate the planet's surface and providing an hospitable home for life.

Once life came on to the scene it too played a part in determining the chemical composition of the atmosphere, Figure 12.2 shows, in a highly schematic form, atmospheric and biospheric evolution over the history of the Earth. Although complex

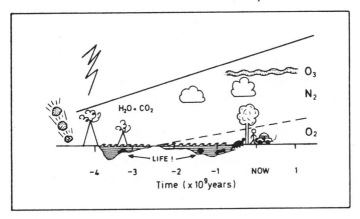

Figure 12.2 *Schematic diagram of the history of the Earth's atmosphere. The influence of the biosphere has been considerable*

life-forms developed only recently, biological processes have combined with physical and chemical mechanisms overall to change the input of 80 per cent water vapour, 20 per cent carbon dioxide and traces of nitrogen and sulphur compounds into the present atmosphere dominated by nitrogen and free oxygen.

Primitive life originated, according to the most widely accepted theories, in the sub-surface layers of the early oceans, where it was protected from ultraviolet radiation. To the first organisms, oxygen was a dangerous poison, a waste product produced from their life process and dissolved in the ocean waters, slowly diffusing into the atmosphere above. The released oxygen was bound up by chemical reactions into the surface rocks, including the iron oxide, "redbeds" formed about 2.6 By ago. But once enough oxygen was around, it changed the rules for living organisms in two ways. First, it provided a new source of energy for life forms that learned the trick of respiration; secondly, by building up in the atmosphere and then forming an ozone layer which absorbed incoming solar ultraviolet radiation, it made it possible for life to emerge from the protecting layers of the ocean.

Enter mankind

Now life could spread on a grand scale, with a dramatic effect on the carbon dioxide content of the atmosphere. Sedimentary deposits buried plants and animals, their remains rich in carbon, in the rocks, resulting in the semi-permanent removal of vast quantities of carbon from the atmosphere. Only semi-permanent, though, since now life, in the form of mankind, is deeply involved in extracting some of that carbon-rich material, in the form of fossil fuels, and burning it.

More subtle variations of biological and atmospheric systems could have produced short-term changes in surface temperature, as when flourishing "tropical" jungles have spread over wide areas of the Earth's land surface and changed the surface albedo and its infrared absorption. One suggestion (J. Lovelock and L. Margulis, *New Scientist*, vol. 65, p. 304) proposes that the atmosphere should be seen as part of the biosphere, producing a complete self-regulating ecosystem which is named after the Mycenaean Earth goddess, Gaia. On this picture, temperatures are confined to a stable narrow band suitable for life by the processes of life itself (Figure 12.3).

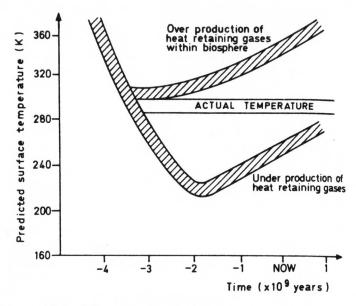

Figure 12.3 *The temperature histories for the Earth under the control of the biosphere. The "Gaia hypothesis" allows temperature control and may have produced the stable temperature curve*

Climate predictions

The two main schools of thought in climate studies give very different points of view. One sees the evolution of the atmosphere as like a wandering walk along a valley floor, with occasional slight excursions up the side of the valley, but soon returning to the most stable conditions at the bottom. The other sees a continuing "ridge walk" balancing in a narrow region of stability with doom of one kind or another but a step away on either side. The role of cyclic changes such as the Milankovitch mechanism which modulates climate at various cycles with periods in the range 10–100 000 years (see *New Scientist*, vol. 75, p. 530, 1977) must also be taken into account, together with evidence that these fluctuations can only modulate glacial epochs when the continents are in the right place – near the poles – for a glacial epoch to occur in the first place (Figure 12.4). So continental drift affects the changing atmosphere!

Other processes, involving more rapid feedback, are still poorly understood. Will increase in carbon dioxide from burning fossil fuels produce a runaway greenhouse effect? Or will a slight temperature increase encourage more clouds to form, increasing the albedo, reflecting solar heat away and cancelling out the

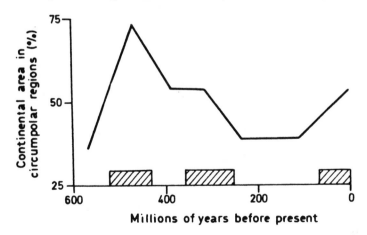

Figure 12.4 *Areas of continental mass close to one or both poles seem to correlate well with (shaded) periods of glacial activity*

greenhouse effect? Predictions at future climatic variations depend first on understanding past evolution of the planetary atmosphere. We are still far from that ideal, but perhaps the current conference in Nice will take us a step further along the road.

26 October, 1978

13

A new theory of atoll formation

HANS HASS

The ring-shaped coral structures of atolls were explained by Darwin in terms of the slow subsidence of an island surrounded by growing reefs. The author's observations in the Maldives lead him to propose, instead, that they are formed from coral reefs which have expanded and collapsed in the centre.

The reason for the curious ring-like form of the reefs that compose coral atolls has caused considerable discussion among scientists ever since Charles Darwin first became intrigued with the problem and made his famous observations during the voyage of the *Beagle* between 1831 and 1836. Subsequent theories have by no means removed the obscurity of the question, and it was thus my intention during the Xarifa Expedition to the Maldive and Nicobar Islands of the Indian Ocean, in 1957–58, to make some underwater observations of the Maldive atolls in the hope that these might further elucidate their origin. For reasons of limited time, however, the new ideas that I put forward here are of a preliminary nature and require substantiation in greater detail by further research work.

Darwin's explanation of atolls was that they started life as fringing reefs round the edges of islands – typically these would be volcanic islands in oceanic areas. The island then subsided, and as the reef continued to grow upwards it became first a barrier reef surrounding the island's centre, and when this eventually sank below the sea, was left as a ring-shaped atoll. But this is by no means the only conceivable process of formation; a second idea, suggested by Sir John Murray after his voyage in the *Challenger* in 1872, is that atolls may have originated on upthrust parts of the ocean floor. Under the right conditions reef-building corals can start to grow at a depth of some 50 metres, and in general, this

Figure 13.1   *Aerial view of one of the Maldive atolls (Credit: Hans Hass)*

would happen on the outside of the sea-mount because of more advantageous growth conditions. Thus a ring would grow upwards to the surface.

Another, more recent, theory due to R. A. Daly invokes the fall in mean sea level which must have accompanied the locking up of vast quantities of water as ice-caps during the different glaciations of the ice age; at the same time the water would have become too cold to permit the continued growth of corals and the existing coral reefs would then have been eroded down to the new sea level. The resulting bare platforms would have provided the bases from which new coral growth began again as the climate turned warmer and the sea rose. Again the maximum growth would be expected on the periphery of the platforms and ring structures would result.

However, the combination of an aerial survey of the Maldives, and some submarine observations of the nature of the reefs, has led me to the idea that at least some atolls may be formed in a completely different manner. Briefly it is that coral islands extend by growing outwards as well as upwards, and that in time their centres subside to form the characteristic lagoons.

Aerial photographs show that all the stages that would occur in such a development are, in fact, illustrated among the Maldives. The first step is represented by isolated cone-like growths of coral which just reach the surface of the sea; next there are bigger islands in which the centres appear to have become sanded over so that corals only continue to grow vigorously on the outer edges; and, as the final stage, one has typical lagoons lying inside circular reefs. Moreover, these beautiful blue lagoons appear to be deeper the greater the diameter of the atoll – the bigger ones are a darker blue.

Thus the idea occurred to me that, as the reefs enlarged, they eventually collapsed in the centre like a badly cooked cake. Subsequent underwater investigation led to a further observation which indeed could well explain by what factors this could come about. While the relatively thin top layer of a coral reef is fairly solid, I found that in the Maldives the main underlying body of the reef has a much less firm structure. These reefs are made up of an open scaffolding of very delicate, highly-branched corals, like

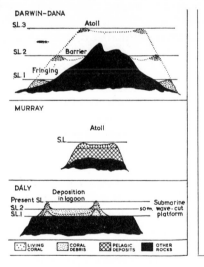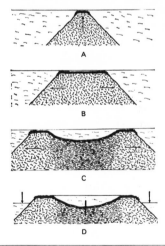

Figure 13.2   *Theories of atoll formation. (Left) three earlier ideas involving relative changes of sea level and the height of the ocean floor (from* Principles of Physical Geography *by F. J. Monkhouse, University of London Press). (Right) Stages illustrating the author's new theory. As the loosely compacted reef extends, it subsides in the centre (A, B, C). This process may be enhanced by trapped tidal waters (D)*

Acropora and *Echinopora,* growing one on top of the other. Broken pieces from dead specimens of corals accumulate as a loose debris slope stretching down into deeper water at an angle of about 45°.

This coral edifice was so loose that I found that I could easily tunnel it in one place, even with bare hands, at a depth of about 20 m. The tunnel soon collapsed, however, because of the instability of its structure. My theory is as follows: after the reef reaches the surface it extends laterally by both coral growth and the accumulation of debris from dead corals. In this way the lattice work of corals that was once on the outside of the reef becomes part of the inside and thus subject to the weight of the accumulated debris lying above it. It seems likely that the calcareous material of this loose scaffolding becomes brittle – it may even recrystallise – and the reef then collapses in the middle due to the weight of material above it. That the subsidence increases with age is borne out by some of the dimensions of the Maldive atolls; 2 km diameter rings have depths between 10 and 20 m, at 8 km the depths are nearer to 40 m, and at 24 km they tend to be about 70 m deep.

The upper crust of the reef is known as the reef flat. It has much greater solidity and strength, largely because it is welded together by shallow water calcareous algae that form a tough crust over the coral fragments. The central subsidence of the atoll does not appear to begin until this platform is a certain size and its centre has become barren of coral growth and sandy.

In the Maldives the different steps illustrating what happens show that the coral reef begins to spread out radially as soon as it reaches the surface and that the corals stop proliferating in its centre, leaving a sort of bald-headed effect. The reason for this is not hard to find and it is, in fact, due to a vicious circle. The peripheral corals have greater access to oxygen and food and tend to starve those in the interior, which gradually die and are ground away to form coral sand. The latter aggravates the deterioration by covering and smothering other corals, over which it is swept by tidal action. In this way even quite small reefs of upwards of 300 yards diameter begin to decay at their centres. It may thus well be that the condition of the reef flat also influences the rate of subsidence by an alteration of its physical properties when it reaches the barren stage.

This would tie in with yet another factor that may affect the subsidence of the atoll's centre. As the reef gets bigger, the length of its rim goes up linearly, but the volume of water enclosed by it goes up by at least the square of its dimensions, even if the depth remains constant. Thus during the tidal ebb and flow the bigger reef circles have to let through a proportionately much greater amount of water than the smaller ones and require more outlets to maintain the same flow. In practice, at ebb tide the water tends to be trapped inside the atoll and, until it can flow out, there is an additional pressure on the bed of the lagoon.

The relative imperviousness of the reef flat would make the process effective in spite of the great porosity of the underlying brittle part of the reef, and tropical rainstorms may also increase the weight of trapped water. It is quite possible that over many thousands of years a rhythmical massage of this kind further encourages the process of central subsidence.

As a final evolution of the atoll, when it grows much bigger, and larger channels intersect the ring, the conditions may again become favourable for corals to grow inside the atoll. It would explain why secondary cone-shaped reefs are seen growing up from the bed of the lagoon. This completes the chain of development which is typical of the Maldives. The young lagoon reefs experience, of course, the same extension and central subsidence, and this accounts for the presence of small rings within the larger ones.

The theory of central subsidence of coral reefs has the advantage that it also satisfactorily explains how barrier reefs originated. As a fringing reef grew progressively farther away from its parent land mass, the resulting subsidence that occurred in its wake would form a deepening lagoon between land and reef. Lagoons of this type are deeper the farther offshore the reef, and the theory of central subsidence, as opposed to those involving geological upheavals and erosion, accommodates this fact just as it does the correlation between diameter and depth of the atolls.

A peculiarity of coral atolls is that they do not occur in all tropical areas. If my theory is correct this could be because not all types of reef are built of the same kinds of coral. Many of them may lack the combination of fragile interior crusted over with a hard reef flat that typifies the Maldives, and gives rise to atolls. Red Sea reefs, for instance, are mainly formed of much more compact corals such as *Porites*. Past theorists may have been misled by the apparent solidity of the upper part of all atoll reefs to assume that the inner parts were equally firm. In the Maldives, at least, this is not the case.

On the other hand, it would be presumptuous to assume that all coral ring structures were formed by central subsidence. Some atolls may be the result of processes like uplift or depression of the ocean floor; others of the ice age mechanism. It is also possible that, in some cases, atolls have been formed by central subsidence combined with one of the other suggested methods. The present theory has the advantage of simplicity in that it does not invoke tectonic or major sea level changes, and explains the origin of atolls and barrier reefs in terms of the normal growth and decay processes of corals.

1 November, 1962

14

Coral timekeepers of the slowing Earth

PETER STUBBS

The fine ring-like banding which is displayed by the skeletons of some corals may represent a daily variation in growth. If so, identical features on fossil corals should provide a powerful method of determining the Earth's rate of rotation in past eras.

A major drawback which plagues the subject of geophysics is that its practitioners have been able to make but few objective measurements that tell them anything about the state of the Earth in former geological times. Thus knowledge of the history of the Earth's development has been essentially speculative and the various theories advanced are highly controversial. It may seem odd that one of the most promising ways to breach this apparently impenetrable obstacle lies in the study of the humble coral and its fossil counterparts. On 25 March a lively interdisciplinary gathering met at the Royal Astronomical Society, to discuss in conjunction with the Geological Society and members of the Zoological Society, the preliminary findings and implications, and the significance, of the growth bands which characterise the outer layers of both some modern and some fossil coral species.

The experimental results, though not easy to obtain, are simple to relate; they are harder to explain. It has been recognised for some time that, like many other marine animals, the calcium carbonate skeletons of many corals show fairly well marked bands representing seasonal variations in their growth which form a tally of the passing of each year just like the growth rings of trees. But there are also other bands which appear to be too close to represent annual growth and which are probably monthly variations related in some way to the lunar cycle. The feature which is now attracting attention, however, and which looks like becoming a potent geophysical tool is the occurrence of very much finer ring-

Figure 14.1 *Fine growth banding on the skeleton of a modern, reef-building coral (Credit: B. R. Rosen/British Museum, National History)*

like banding in some corals. Typically these structures can be seen preserved on the epitheca, the external layer on the lower part of the coral's conical skeleton. They may number between 20 and 60 per mm (see Figure 14.1). If the first and second order banding represent, respectively, yearly and monthly growth fluctuations, it is a reasonable guess that these finer bands are also connected in some way with time. The question is, are they daily variations, tidal ones, circadian (having a built-in biological clock with an approximately 24-hour periodicity) or what?

The exciting aspect of the work is that if these delicate responses of corals can be proved to represent daily variations, then counting how many occur between each annual band gives a possible way of finding out how many days in the year there were at various times during the development of the Earth. It is an established fact that the length of the day has been changing as a result of the tidal friction produced in the oceans by the Moon, by something like two seconds every 100 000 years. If this rate has been constant over geological time it would mean that there were about 428 days in a year at the beginning of the Cambrian Period – 570 million years ago – and rather more than 400 days per year in the Middle Devonian Period – 370 million years ago.

In addition, however, to the external effects of the Moon's drag, which result in a transfer of angular momentum from the Earth to the Moon (so that the Moon, as a result, has continued to get farther and farther from the Earth), the length of the day could also have been affected by internal processes that produced a change in the Earth's moment of inertia. The Earth cannot change its angular momentum except by transfer to the Moon, and physical laws insist that the total angular momentum of the virtually isolated Earth–Moon system must remain fixed. But the moment of inertia of the Earth – a property of its shape, size and the distribution of matter inside it – could have altered during the evolution of the Earth either if the Earth had expanded or contracted, or if its dense core had been gradually enlarging with the passage of geological time by the diffusion of iron, and possible other heavy elements, inwards. As Professor S. K. Runcorn of the University of Newcastle upon Tyne explained, if we also know how many days there were in the ancient month it is possible by taking Kepler's laws of planetary motion into account to separate changes in the Earth's rate of rotation caused by tidal friction from those produced by factors that have altered the Earth's moment of inertia. In this way one can obtain values for the different moments of inertia at particular times in the geological past and compare them with the various planetary theories that have been advanced.

A number of workers are now busy seeking suitable corals, counting their bands and trying to devise more objective ways of recording the growth fluctuations. One of the best established results obtained so far is that from Middle Devonian corals collected from New York and Ontario. The annual growth rates counted by Professor J. W. Wells of Cornell University vary from specimen to specimen, even where the absence of damage to these fragile structures permits a full record, but all are greater than 365. The range lies between 385 and 405, and gives an average value of close to 400 in accordance with the computations, on tidal friction (changes in the Earth's moment of inertia would appear as a smaller effect and present results do not seem to be accurate enough to detect these unequivocally). Professor Wells also gets a smaller figure of 380 days for the length of the younger Carboniferous year. One older Ordovician specimen gave an anomalously large value and it is clear that many more data are called for before firm conclusions can be reached. It is gratifying, though, that he has found that modern corals from the Caribbean appear to have something close to 360 bands to the supposed year.

As far as determinations of the monthly aspect goes – that is whether one can establish that the second-order banding really represents lunar monthly fluctuations – Dr C. T. Scrutton, of the British Museum (Natural History) has done some revealing work on 10 Middle Devonian specimens from North America. He gets an average of 30.6 third-order ridges to every second-order band. To test that this could represent the number of days in a Devonian month, he divides this figure into 399 – the length of the Devonian year on the basis of tidal friction computations. The quotient comes out to 13.04 which looks suspiciously like the number of lunar months in the year (at present there are nearly 12.4 (synodic) lunar months in the year).

Before the geophysicists embark on a major fossil-hunting and coral-counting programme, however, they want to be sure that there is some biological justification for their assumptions about the growth bands. It appears that coral biology is a relatively neglected subject, partly because corals are rather temperamental animals to keep in captivity. As yet, for instance, no one has done what appears to be the obvious experiment of watching a coral grow to see if it produces one fresh band every day, with a thicker one each lunar month and an overall annual waxing and waning of its skeleton. Some work in this direction has now started, according to Dr T. F. Goreau, at the University of the West Indies.

The annual banding is the simplest to explain and is almost certainly due to seasonal variation in the amount of calcium carbonate which the organism deposits in its skeleton in accordance with overall temperature changes. The "monthly" and "daily" growth bands are not so self-evidently accounted for. If the monthly bands are the result of some tidal effect upon growth one might expect a twice-monthly set of growth constrictions corresponding to the bi-monthly spring tides. Dr Scrutton believes his monthly bands are most plausibly explained by a lunar breeding periodicity which is known to occur in some modern corals which stick closely to the lunar cycle.

When we come to the fine-scale banding the problem becomes more acute. The banding shows no signs apparently of being a double feature due to the 12-hourly tidal fluctuations. If it is a genuinely daily phenomenon the obvious cause to seek is one which relies upon the alternation of light and dark. There are, however, two rather different kinds of corals with different habits. One kind, the reef-building corals, as Dr C. M. Yonge of Glasgow University explained, live in a symbiotic relationship with numerous algae which inhabit their tissues and, as a result of the photosynthesis which they perform, supply the coral with oxygen, removing unwanted phosphate, ammonium and carbon dioxide. They do not themselves constitute food for the organism. Such corals do indeed display a diurnal difference in the amount of calcium carbonate which they take up and which may fall by as much as 20 per cent at night. According to Dr Goreau the deposition of this calcium carbonate as skeletal material is a complex metabolic process that could well be controlled by the removal of carbon dioxide by the algae. This, in turn, would be controlled by the amount of light available for photosynthesis.

There is, however, one very large snag in that the second group of corals are not reef builders at all; instead they live at great depths (*Flabellum* has been found 3300 m down), many of them in total darkness and a uniform temperature. They do not possess symbiotic algae, yet some of them show all the three kinds of banding (Figures 14.1 and 14.2). How are we to explain this? Light can have nothing to do with it.

Speculating, Dr Goreau suggested that the answer might lie in feeding periodicity. The electron microscope indicates structures in the outer cells of corals that could imply a subsidiary method of absorptive feeding. Radioactively labelled amino acids seemed to be taken up by experimental corals and he proposed that, rather than responding to light, the corals might be responding to the *smell* of plankton, the small marine organisms on which they live. Plankton rise and fall daily to and from the water's surface and this rhythm may account for the skeletal growth fluctuations. But

Figure 14.2 *A deep-water coral*, Flabellum curvatum. *Fine growth bands are evident despite the dark habitat (Credit: J. D. Taylor/British Museum, Natural History)*

plankton do not sink below a certain depth, also because of light requirements. Below this, Dr Goreau asks, could the absorptive subsidiary feeding mechanism come into play? Carbon compounds dissolved in sea water can apparently be precipitated as particles by the action of bubbling and, presumably also by tidal surf. Do waves of these nutritional particles "snow down" after every high tide to keep the deep water corals supplied? And is there any chance that the periodicity is preserved down to these great depths? It seems likely at any rate that particle size distributions would put paid to any possibility of getting an exact tidal record, preserved in the coral skeletons. Alternatively, are we dealing with inbuilt biological clocks which could well be independent of outside influences?

Clearly more work is called for and better ways of assessing the band counts. Already at the University of Newcastle upon Tyne, Dr K. M. Creer is attempting to count them with a microprobe analyser by detecting subtle chemical variations that occur between one band and the next; while Professor Runcorn suggests that photographs of mechanically recorded profiles could be used as diffraction gratings to obtain spectra of the frequencies present on these marine relics – truly a hybridisation of the sciences.

31 March, 1966

15

A Precambrian nuclear reactor

HOWARD BRABYN

French investigations, reported at a recent scientific meeting, revealed that a natural nuclear reactor established itself long ago in the rocks of West Africa. If the phenomenon is widespread, this fascinating piece of geology could spell difficulties for the uranium trade.

When the first self-sustained, controlled nucler reaction took place on a Chicago tennis court on 2 December, 1942, it seemed that man had finally surpassed nature, unlocking the door to a vast new energy source which nature herself had left untried. Yet 1700 million years ago in a sub-equatorial region of present day Gabon in west Africa, natural forces built a water-cooled nuclear reactor that operated intermittently for as long as one million years. This was the astounding conclusion reached by nuclear physicists of the French Commissariat a l'Energie Atomique (CEA) as revealed in two papers presented this autumn to the French Academy of Sciences by Dr Francis Perrin, former chairman of the CEA (see Monitor, *New Scientist,* vol. 56, p. 6).

For the first time, untreated deposits of uranium, at Oklo in Gabon, have been found to contain less than the hitherto invariable 0.72 per cent of the isotope uranium-235. The only possible explanation for this phenomenon seems to be the occurrence of a spontaneous nuclear reaction thousands of millions of years ago before uranium had declined from a "naturally enriched" state through the process of radioactive decay.

Was the discovery at Oklo a unique freak of nature, a scientific curiosity of immense theoretical interest but little practical consequences? Or will the specialists, sent scurrying by Dr Perrin's announcement to make new assays of uranium deposits through-

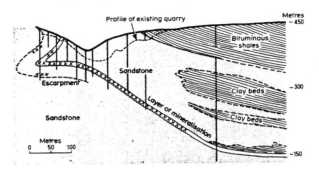

Figure 15.1 *A cross-section through the Oklo quarry showing the various "components" of the fossil reactor*

out the world, find the story repeated elsewhere? If they do, of course, it will have considerable repercussions upon future energy policies and the economies of the uranium-producing countries.

The first link in the chain leading to the discovery of the fossil reactor came on 15 June this year when a specialist at the CEA's production control laboratories detected an abnormally low uranium-235 content in the uranium hexafluoride produced at the Pierrelatte enrichment plant from mineral concentrates treated in France.

Follow-up analyses conducted by CEA experts rapidly eliminated the first hypothesis, namely that uranium already used in a reactor had accidentally become mixed with the natural uranium. The anomalous material was eventually traced back to a particular section of the Oklo deposit in the Haut Ogoué region of Gabon which had been worked since 1970 by the Compagnie des Mines d'Uranium de Franceville (COMUF). At the nearby site of Mounana, in operation since 1961, COMUF has built an initial treatment plant at which the extracted ores are pre-concentrated prior to shipment to France for purification and enrichment.

Isotopic analyses indicated that the mixture of ores from Mounana and Oklo gave the same abnormally low concentrations of uranium-235 as those discovered in pre-concentrates delivered to France; and that the richer the ore in uranium the lower was the content of uranium-235. Concentrations as low as 0.44 per cent were discovered, but this was not the only surprise; some samples in fact showed slight enrichment, in some cases attaining 0.74 per cent.

This last factor tended to confirm the spontaneous nuclear

reaction theory, the argument being that, just as in a man-made reactor, the chain reaction had produced a quantity of plutonium. Over the years this had decayed back into uranium which was younger and therefore richer in uranium-235 than primordial uranium. Perhaps more compelling backing for the hypothesis came with the discovery of four rare elements in a sample from Oklo – neodymium, samarium, europium and cerium – with isotopic compositions hitherto found only as a result of man-made nuclear reactions.

The French nuclear scientists realised that if they could show that the appropriate conditions had been fulfilled, the evidence for a spontaneous nuclear chain reaction, occurring millions of years ago when the natural content of uranium-235 in uranium was about 3 per cent, would be overwhelming.

To satisfy those requirements the concentration of uranium would have to be high; some form of moderator and coolant would be required; and the area would have to be free of neutron-absorbing elements such as boron or cadmium which would have inhibited a chain reaction. Geological studies show that at the Oklo of the Precambrian era all these conditions were met.

The uranium-bearing deposits in Gabon lie some 38 miles west-north-east of the town of Franceville. They are part of a series of sedimentary rocks which 1700 million years ago filled up vast depressions in the granitic and crystalline basement of the region. Gabon is a country of abundant rain and the hydrological and climatic conditions at the time played an important part in determining the nature of the sedimentation and also in the mineral accumulation. Sedimentation in the Franceville basin followed a sequential pattern in which an initial deposit of conglomerates was followed by finer deposits ranging from sandstone to silt. Mineralisation occurred during the first part of the sedimentary cycle between the deposition of sandstone and finer, impermeable sediments. After they had been deposited these sediments were subjected to powerful tectonic upheavals which caused important transformations in the basin and played a large part in the reconcentration of the uranium-bearing ores. The deposits, in fact, border a fault system running north-west between a crystalline rock formation and the Franceville sediment-ary basin.

All these mineralogical and geological factors add up to ideal conditions for the occurrence of a spontaneous nuclear reaction – a high concentration of fissile material (even today in some sectors

Figure 15.2   *The eastern side of the Oklo quarry seen from its western edge (Credit: CEA)*

the uranium content of the ore mined at Oklo exceeds 10 per cent), lying in a 20-foot layer of water-saturated sandstone beneath a layer of impermeable clay, in an area almost totally devoid of neutron-absorbing elements. Nature had, in fact, painstakingly constructed her own heterogeneous, enriched uranium nuclear reactor (see Figure 15.2).

The indications are that once ignition occurred the reaction continued until water permeating the layers of sandstone and uranium turned to steam. It then halted to allow cooling, only to break out again in another sector. Like some giant witch's cauldron the whole Oklo site simmered and spluttered for nigh on a million years.

This extraordinary discovery is, of course, of great intrinsic interest, but its practical consequences are as yet difficult to assess. Should it turn out to be an isolated phenomenon, it will no doubt be something of an economic blow to Gabon, but it will have only marginal effect on total world reserves of uranium. Buyers of uranium ore, however, are now likely to be more wary in their purchasing and may well demand an "isotopic" quality guarantee when placing forward orders. If the same sequence of events has been repeated many times elsewhere, then it may be necessary to make a new, more critical assessment of world stocks of fissile material.

One thing is certain, the process of radioactive decay ensures that no such fantastic combination of circumstances can ever occur on our planet again.

9 November, 1972

16

Putting an age to the fossil reactor

Recent French studies reveal that the curious "fossil reactor" discovered in Gabon in 1972 was active some 1780 million years ago. Research on the isotope abundances in the uranium deposit where this strange phenomenon took place confirm that a natural, moderated, fission process did indeed happen in these strata at Oklo, as the French scientists of the Commissariat a l'Energie Atomique had reported.

Joel Lancelot, Annie Vitrac, and Claude Allegre of the Institut de Physique du Globe, Paris, found that both the total uranium content, and the ratio of uranium-238 to uranium-235, were unusually high and also very variable in the ore body. The anomalies are only easily explained on the fossil reactor hypothesis.

However, considerable secondary changes in the abundances appear to have occurred since; for, had the reactor just ceased its operation and left its products behind unaltered, there should now be a fixed relationship between the total uranium amounts and the $^{238}U/^{235}U$ ratios. The French trio found abnormal uranium ratios for all ore samples containing more than 40 000 parts per million of the element. Below this figure the isotope ratios were normal suggesting both that the richer ore had constituted a critical mass of enriched uranium, and also that the "spent fuel" had subsequently been dispersed by some agent (*Earth and Planetary Science Letters*, vol. 25, p. 189).

Presumably water filled the role of moderator to produce the thermal neutrons in the reactor. This model is borne out by the age measurements from uranium/lead ratios. These imply that the reactor went critical around the time of its inclusion in the sediments that enclose it. The three Paris workers, using the ore with "normal" isotopic composition, have placed the age at 1780

±60 million years ago, well back in Precambrian times. That age agrees well with a rubidium/strontium age for the surrounding Franceville formation found by another researcher. During the settling of the sediment water would have been present to perform the task of moderating the reactor.

"MONITOR", 27 March, 1975

PART THREE
The Revolution in Earth Sciences

17

Rock magnetism and the movement of continents

PETER STUBBS

For half a century geologists have argued about the theory of the Austrian meteorologist Wegener – that the present continents have, in former ages, drifted about on the Earth's surface. The study of rock magnetism during the past ten years has virtually settled the issue.

Rocks with a natural magnetism have been known to mankind since the Chinese discovered many centuries ago that suspended pieces of lodestone would serve as compasses. Pure lodestone is a fairly rare mineral, however, and it was the explorer Humboldt who in 1794 recorded one of the earliest instances of magnetism in a more ordinary kind of rock, serpentine. His observations remained undeveloped until the beginning of this century when some sporadic, although interesting, experiments began to be made in a more systematic fashion. During the past decade the subject has gained considerable importance and is now one of internationally wide appeal to geophysicists.

The basis for rock-magnetic studies lies in the assumption that the magnetisation which can be measured in certain types of rocks represents accurately the direction of the Earth's magnetic field in former geological eras. The rock-magnetist is then able both to deduce facts about the past behaviour of the Earth's field, and also to investigate the highly controversial idea that the continents have moved about over the Earth's surface in the course of their long history.

Rocks become magnetised in different ways according to their nature. Igneous, or crystalline, rocks solidify from a molten condition at a high temperature. Many of them, particularly the ones described as "basic" – for example basalt or dolerite – contain small grains of the iron oxide magnetite, chemically the same mineral as lodestone. As they cool they reach a certain critical temperature called the Curie Point (after its discoverer, Pierre Curie) at which they acquire a relatively strong and highly stable magnetic polarisation in the direction of the surrounding magnetic field.

Sedimentary rocks, on the other hand, made up of layers of particles ground away from pre-existent rocks, become magnetic in an entirely different way. In some of them sparsely distributed magnetite fragments, already magnetised during cooling at an earlier stage, become aligned in the Earth's magnetic field while they are settling through water: in fact they behave like small, freely suspended compass needles. Most of the rock-magnetic results from sedimentary rocks, however, have come from red

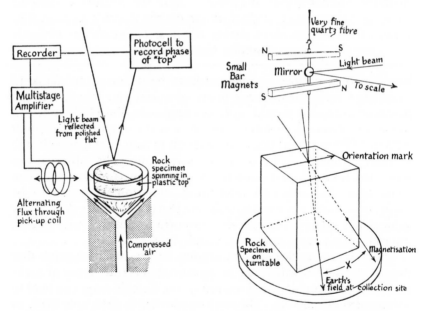

Figure 17.1   *The two common methods of measuring the direction and strength of magnetism in a rock. In the astatic magnetometer method, on the right, the specimen is rotated into successive positions below the magnet system, and the angular deflection of the latter observed each time on a scale. In the second technique a spinning specimen induces a weak alternating current in a coil. This is amplified, and its image compared with that (measured optically) of a reference mark on the "top"*

sandstones and mudstones, formed typically under continental rather than marine conditions. The red oxides of iron, such as haematite, which supply the colour, also seem to provide the magnetism. It is likely that magnetisation is a chemical process associated with the oxidising conditions under which such rocks accumulated, though the matter needs further investigation. The magnetism of sedimentary rocks is very weak by ordinary laboratory standards and very sensitive equipment is needed to measure its directions and intensities (see Figure 17.1).

What do the results of measurements like this tell us? First of all let us consider the present shape of the Earth's magnetic field. It can be analysed into two parts. The major one is what is called the axial dipole field, and is described by a configuration of imaginary lines of magnetic force running through space from the south to the north pole. The lines are exactly those which a bar-magnet, placed at the Earth's centre, would produce if it were directed along the Earth's rotational axis. It is something like the pattern one gets from sprinkling iron filings on a piece of paper above a bar-magnet. The lines of force incline towards the Earth's surface at different angles of dip, related by a simple equation to the latitude at which they are measured. They are horizontal at the equator and vertical at the poles.

The other, smaller component of the Earth's field, the secular variation, changes slowly with time, having a cycle of about a thousand years and imposing discrepancies on the dipole field of up to 20 or 30 degrees in direction. Averaged over periods of the order of a thousand years the total field approximates closely to an axial dipole one.

The earlier workers in rock magnetism concerned themselves with measurements on rocks of younger geological periods in an attempt to trace the pattern of the secular variation in past ages. While their studies in this branch of the subject did not give a detailed picture, they nevertheless established that the Earth's field behaved much as it does now back for about the last 50 million years approximating to an axial dipole type with a scatter in the results which would be expected from similar secular changes. Careful work on the magnetism of archaeologically datable pottery kilns and their contents, notably by Professor E. Thellier at Paris, is now yielding accurate values of the secular variation in historic times.

An early discovery of startling unexpectedness, however, was that about half the rocks measured were magnetised in exactly the reverse direction to the present field. The phenomenon extends to older rocks as well, and it now seems highly likely that the Earth's magnetic field has reversed itself repeatedly in the geological past, although perhaps not very regularly. This has far-reaching implications for theories as to why the Earth has a magnetic field, and the eventual pattern of reversals may tell us more about its cause. Current ideas favour some kind of self-stimulating dynamo set up by fluid motions in the Earth's liquid, and almost certainly conducting, core. It is possible to envisage a dynamo system of this type capable of undergoing intermittent current reversals.

When rock-magnetists began to look at the older geological formations a further surprise occurred. In 1952 E. Irving, at Cambridge, measured samples of the 700–800 million-year-old Torridonian Sandstone of north-western Scotland, and a group at Manchester University under Professor P. M. S. Blackett and Dr J. A. Clegg examined the 180 million-year-old Triassic Sandstones of Cheshire. Both discovered directions of magnetisation which, instead of being roughly parallel to the present field at the collection sites, diverged very considerably from it. In particular the angles of magnetic dip disagreed with those of the present field.

Geophysicists at once realised that such results could well be interpreted as meaning that Britain formerly occupied a substantially different latitude, or alternatively that the Earth's magnetic pole had wandered across its surface. The opportunity to resolve either of the long-standing bones of contention about continental drift or "polar wandering" encouraged a wave of rock-magnetic exploration extending over most of the world, which is continuing strongly.

At the present time well over a hundred similar geological formations having divergent magnetisations have been found, ranging back in age over the past 2000 million years. We must remember, though, that only rocks younger than 500 million years can be dated with reasonable accuracy. (The oldest rocks known go back to about 3500 million years.)

Most of the data come from north-west Europe, North America, Australia, India and South Africa. Work is also progressing, however, in South America, which forms part of the programme of a research group at King's College Newcastle-upon-Tyne, under the direction of Professor S. K. Runcorn, and also in Russia, Japanese scientists, who were in the forefront of the early rock-magnetic work, are limited by Japan's geology to a range of younger rocks but have done a lot of valuable analysis of the solid-state magnetic properties of rocks. A team at Birmingham University has recently produced some of the first measurements on rocks from Antartica.

The pros and cons of the continental drift hypothesis have had a very stormy history of dispute since the idea was first propounded by the Austrian meteorologist, Alfred Wegener, in 1910 and it is as well to proceed cautiously in assigning this explanation to the findings of rock-magnetism. What other natural processes could conceivably have the same effect?

It was with this question in mind that Professor Blackett, Dr Clegg and myself recently made a collective survey of the results of rock magnetism (*Proceedings of the Royal Society A,* vol. 256, p. 291). The object was to define statistically just what it was that scientists had found out, making as few assumptions about its eventual explanation as possible, and then to discuss all its possible interpretations. To be able to use a single statistical quantity, we calculated the total angle – the divergence – between the magnetic polarisation of each given rock formation and the direction of the Earth's present dipole field at the relevant locality (angle X in Figure 17.1). By plotting a graph of the divergence against the geological age of the formation, one can see at once that the angle increases progressively with age in a statistically consistent and non-random fashion. That is, the older the rock the further removed is its magnetisation from the present field direction. The result holds true for each of the land masses of Europe, North America, Australia and India for which the data are most substantial. It represents the incontrovertible phenomenon which this branch of rock-magnetism has revealed. How do we explain it?

One possibility is that the rocks did not take up, or did not preserve, a magnetism in a direction parallel to the field prevailing when they were formed. Gravitational effects like compaction, or mountain building or the weight of overlying rock, might be invoked to justify such a view. The objection is that none of these agents is competent to produce a *progressive* change in the divergent angle. Also such effects would certainly affect igneous and sedimentary rocks to a different extent. In fact the results for different rock types are in good agreement.

Another conceivable interpretation is "polar-wandering", by which we mean a sliding movement of the Earth's crust as a rigid unit relative to the magnetic poles. One way of looking at rock-magnetic results is to represent each by a suitable past position of

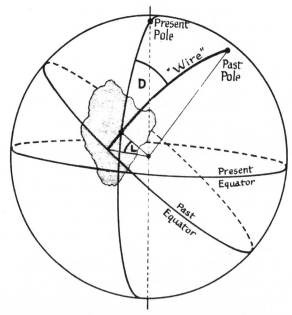

Figure 17.2 *How a past pole position relative to a continent is computed. Angle D is the measured declination. Angle L is the ancient latitude of the rock collection site, given by tan I = 2 tan L, where angle 1 is the measured angle of mangnetic dip*

the pole. What we measure is the dip, or inclination, of the fossil magnetism, corresponding to a past magnetic latitude of the particular continent, and the bearing, or declination, of the pole from the land-mass. We can make a global model by way of illustration (Figure 17.2), and for each geological period, attach to each continent a piece of wire with a pole on its outer end. The inclination tells us the length of the wire, and the declination its bearing with respect to the continent.

Obviously if polar wandering is the sole explanation of the phenomenon then, at any particular time, the poles from the different continents should all coincide. In fact, they do not, and although polar wandering may have occurred, alone it cannot provide the whole answer.

A third suggestion is that the Earth's field in the past may not have been like that of an axial dipole at all, but something with a more complex configuration which changed very slowly with time.

Although this theory cannot be disproved, there are strong theoretical arguments to support the axial dipole hypothesis, and we have already seen that the field has continued in such a pattern for the past 50 million years or so. In due course, measurements on rocks of one age, dispersed over a single continent, may clear up this unlikely possibility.

Finally, then, we are forced back to accepting continental drift – relative movements between the land-masses – as the most plausible reason for the divergent magnetisations.

Many people, in addition to Wegener and the American F. B. Taylor, have developed theories on drift, notably the South African geologist A. L. du Toit. Geological evidence to support the idea is based upon such features as matching coast lines, and "tie-ups" between continents formerly alleged to have been connected together. Tie-ups may be correspondences of mountain structures, sedimentary sequences or of fossil distributions, all pre-dating the separation of the land-masses. The need to find origins – source lands – for enormous basins of sediment in places now covered by oceans is a further argument. Perhaps the strongest line of attack is the climatological one. The well-known instance of the Permo-Carboniferous climate (about 200 million years ago) is among the more convincing pieces of evidence. During this time, as Figure 17.3 shows, large low-lying areas of South America, South Africa, India and Australia were glaciated. No corresponding ice-cap existed in the northern hemisphere and the most likely explanation is that these continents were grouped more closely together about

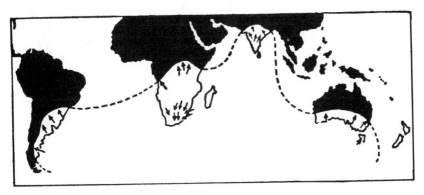

Figure 17.3 *Map showing the extent of the Permo-Carboniferous glaciation – arrows indicate the directions of ice-flow*

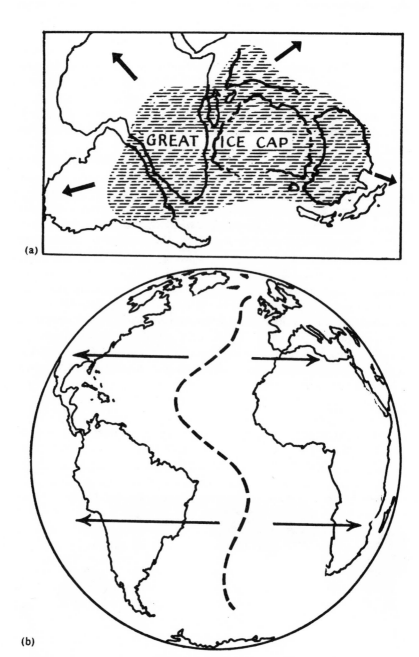

Figure 17.4 (a)   *Radial drift of southern hemisphere continents from the glaciated South Pole*
Figure 17.4 (b)   *Shows westerly drift of the Americas from Europe and Africa opening the Atlantic Ocean*

the South Pole. The picture is enhanced by the fact that at the same time, tropical coal forests extended across a now temperate zone from North America, through northern Europe to Russia.

Common to most detailed theories of continental drift is the idea that there were formerly two super-continents in the northern and southern hemispheres. When they broke up the general pattern of movement included a radial shift of the southern hemisphere continents away from the South Pole, coupled to a westerly displacement of North and South America away from Europe and Africa (Figure 17.4). How does the rock-magnetic picture of events fit the geological one?

If we assume, for simplicity, no polar-wandering, and merely consider the total independent movement of each continent relative to a fixed pole, we can produce the summary of rock magnetic measurements shown in Figure 17.5. From the magnetic inclinations and declinations in the rocks we calculate the latitude of some reference point on each land-mass during a particular past period. The full circles on the map are placed at these latitudes, the number attached to each representing the age in millions of years when the reference point occupied the position. The arrows,

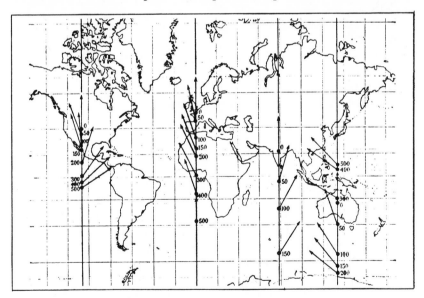

Figure 17.5 *Summary of rock magnetic data. The number against each point is the corresponding age in millions of years*

derived from the measured declinations, show how the continents have rotated, each arrow corresponding to the past position of an imaginary line on the continent at present running due north.

Figure 17.5 tells us, in general, all the kind of information we can glean from rock magnetism about continental drift. The details of the picture will be filled in by subsequent work, but, for instance, we shall be able to deduce nothing directly about the past longitudes of the continents. On the map past positions of reference points are shown fixed to arbitrary meridian which are the present longitudes of the points. Thus we cannot test in a straightforward manner the westerly drift of the Americas.

Nevertheless, the general northward, radial drift of the continents away from the South Pole is very striking and fits well with Wegener's original ideas and with the evidence of past climates. North America seems to have rotated in an anti-clockwise direction a total of about 50° relative to Europe. This refutes some of the particular kinds of fit which geologists envisaged. Australia has behaved in a very interesting fashion, sliding first southwards to the Antarctic and then northwards. Despite such eccentricity, this double movement receives excellent support from the climatic evidence.

As rock-magnetic results accumulate and are supplemented by further extensions of palaeoclimatology, continental drift will almost certainly come to be a reputable concept, perhaps assisting applied geology in the search for oil, minerals and so on. Scientists are becoming convinced in greater numbers that there really may be some truth in this much disputed theory.

1 June, 1961

THE ORIGIN OF OCEANS AND CONTINENTS

There is a growing interest in the Earth's mantle – the 2000 mile layer that forms the bulk of the Earth between the thin crust and the liquid core. An international committee was set up at Helsinki last year to coordinate research into the upper mantle. Already the simple picture proposed by early earthquake seismologists has been complicated by the late Professor Gutenberg's low velocity layer. In a recent article in *Nature* Dr R. S. Dietz puts forward a new model of mantle behaviour. The basic cause of all the forces that have shaped the Earth's surface into continental and submarine mountains and valleys is ascribed to convection currents in the mantle.

It is not a new idea, but the proposed consequences of such convection, in Dr Dietz's excellent exposition, go a long way to weld together many of the new facts which have been collected by the modern schools of oceanography. The upwelling parts of the convection currents mostly manifest themselves in the oceans, and the flow patterns spread side-ways and circulate downwards some thousands of kilometres away. The zones of downward movement are associated with the deep ocean trenches or with the continents. The floor of the ocean is slowly creeping outwards from the upwelling zones at a rate of about an inch a year.

Such continual spreading of the sea-floor explains the rift valleys which have been mapped in all the oceans and have demonstrated the existence of the tensions in the Earth's upper layers. At the same time it fits nicely with the compression forces that are needed to account for the buckled rocks which form mountain ranges on the continents. The continents are like giant icebergs about 35 km thick, floating on the sea of plastic mantle. They are pushed into position by the convection currents, and come to rest about zones of downward movement. The old concept of wandering continents is readily acceptable on this basis. The Atlantic Ocean is the result of a convection which has upwelled in the mid-Atlantic Rise and has pushed the Americas away from Europe and Africa.

The thin crustal layer which has been shown by seismic measurements to cover the sea floor must now be regarded as merely the upper part of the mantle. It is not different chemically, but is merely a low-pressure form of the mantle rock. It may well be rejuvenated to its mantle form when it reaches the downward part of its continuous circulation. The few thousand feet of sea-bed sediments may be scraped off as the circulating sea-bed pushes down the flanks of the continents, and may thus be adding to the continental material. That would be why no very old sea-bed sediments have been found – they have been swept tidily under the continental carpets. It is also apparent why the lighter crustal rocks are not spread in one uniform layer over the whole surface of the Earth, instead of being accumulated as continents.

Many more facts will be needed to corroborate this synthesis of geological, seismological and oceanographical discoveries. Not least will be the need for chemists to look into the changes of mantle rock that can take place under various conditions of temperature and pressure.

"MONITOR", 29 June, 1961

18

The origin of continents

A. E. SCHEIDEGGER

Geophysical evidence suggests that the Earth's land-masses are structurally distinct from the ocean floor. The primeval parts of the continents are thousands of millions of years old, and the mechanism of their origin is highly speculative. The Mohole project may solve many of the questions posed in this article.

The existence of continents is a remarkable feature of the Earth's surface. If our globe were perfectly homogeneous and symmetrical, water would cover the surface to a considerable depth, and human life as we know it would be impossible. As it is, continents represent only about one-third of the surface. regarding our globe as a whole, we must therefore admit that the oceans are the norm, continents the exception. the questions arise: do the continents differ from the ocean bottom in their basic structure, or are they merely parts of it which have been uplifted? If the latter is the case, can they rise and disappear in tremendous cataclysms? The thought of such upheavals has fascinated man for centuries, as witness the myths of sunken continents such as Atlantis.

As recently as twenty years ago the question concerning basic structure could not have been answered with any degree of certainty. However, recent oceanographic expeditions have definitely established that the upper parts of the Earth beneath oceans are significantly different from those beneath continents. Over the whole globe one finds a surface at various depths at which the velocity of sound waves (such as of waves created by explosions or earthquakes) jumps from a low value to a very high one. The discontinuity was discovered by A. Mohorovicic of Sagreb, Yugoslavia, and has been called after him. It lies at an average depth of some 30 km beneath continents, but at a depth of only 5 km beneath the ocean bottom. Geophysicists call everything above the Mohorovicic Discontinuity the "crust" of the Earth, and hence they say that below continents the crust is roughly 30 km thick, but below oceans only 5 km thick.

It is the aim of the American scientists working on project Mohole, of course, to drill down to the Discontinuity and reveal its true nature. At present there is much speculation as to its significance. Most scientists assume with Professors H. Jeffreys or A. Holmes that it represents the boundary between two chemically different substances such as the granite-like rocks on the one hand and the olivine type, such as dunite, on the other. Others, however, such as G. Kennedy, think that it is merely a physical change brought about by pressure – like that between diamond and graphite, both of which are chemically carbon. J. T. Wilson has recently put forward the idea that the Mohorovicic Discontinuity may be the Earth's early surface, corresponding to what we see on the Moon at present.

The recognition of the Mohorovicic Discontinuity, and its mapping over most of the globe, has been one of the most important recent achievements in the Earth sciences. It shows that there is a basic difference between continents and oceans. Even if the Discontinuity represents only a phase change rather than a discontinuity of composition, the fact that it lies so much deeper beneath continents than beneath oceans retains some significance.

Continents are very old features of the Earth. In the oldest parts of the continents (the so-called "shield" areas) rock ages of around three thousand million years have been determined by recent radioactive dating, for instance by the method which involves decay of rubidium-87 to strontium-87. Thus, the creation of continents dates back to the Earth's early history, and is probably linked with its origin about 4.5 thousand million years ago.

One generally distinguishes between two types of theories of the origin of the Earth, the *uniformitarian* and the *cataclysmic* theories. The *uniformitarian* theories, which go back to Kant and Laplace, in the 18th-century, assume that the Earth evolved slowly, growing by the condensation of material from the Solar System. Scientists have envisaged the manner in which this happened in a variety of ways. For instance, T. C. Chamberlain in 1921 favoured the cold accretion of interstellar material and C. F. von Weizsäcker in 1944 elaborated a theory of the collection of dust at the boundaries of vortices.

Contrariwise, in the *cataclysmic* theories it is assumed that the Solar System was created by some sort of catastrophe. Professor

Fred Hoyle assumed that the Sun was part of a binary system and that the Sun's companion blew up as a supernova and the hot gas condensed to form planets. Similarly, Sir Harold Jeffreys, like Sir James Jeans, assumed in 1929 that the Sun, originally bigger than it is now, was disrupted by tidal resonance set up by the gravitational pull of a passing star.

We do not know whether the Earth started as a *cold* or as a *hot* body; yet, when we are seeking to determine how continents originated, this is a crucial question. We cannot in fact expect the final answer to our quest until some very fundamental problems concerning the origin of the Earth itself have been settled. The present limitations of our knowledge make it unlikely that this can be achieved in the near future. All that one can do at the present time is to consider the various possibilities and their implications.*

The original continents may have been vastly different in size and distribution from the present ones. Wegener's theory that they have drifted about on the surface of the Earth, propounded in 1910, has received much support within the past ten years from studies of rock magnetism. Certain geological observations indicate also that the continents have *grown*. For one thing, the oldest rocks are invariably found in the centres of continents, in the shield areas. Thus, these areas can be regarded as nuclei from which continents have grown outward by the addition of material both through successive phases of sedimentation followed by mountain-building, and also by the addition of volcanic material (Figure 18.1). It is possible that the bulk of the continents is formed in this way by the eventual re-crystallisation of sediments derived from lavas. J. T. Wilson envisages the process as one of continuous exudation of lava flows from advancing zones of fracture in the Earth's crust.

Although such a picture seems well established, it should be remarked, for the sake of fairness, that not all scientists accept it. On some occasions, also, old rocks have been found where they do not seem to belong (such as in California), and for this reason some geophysicists contend that the continents have never grown much. It is to be hoped that this particular issue will be settled as, increasingly, the age of rocks is determined by modern physical methods.

---

*For the reader interested in greater detail it may be noted that the author has reviewed theories discussed here in his book *Principles of Geodynamics* (Springer, Berlin 1958).

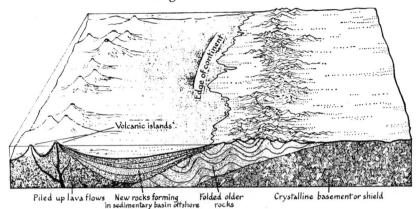

Figure 18.1 *The various ways in which a continent may grow, shown schematically*

In considering how the continents may have been created, we must keep in mind that, as noted above, their present distribution over the surface of the Earth does not necessarily tally with their original position. Most theories of continental evolution assume a different distribution of continental material over the surface of the globe from that which we see today.

There are, however, also theories which assume that there has been relatively little movement, and that the present distribution of continental nuclei has persisted since primeval times. They base their deductions on the observation that four of the important old continental shields are located roughly at the corners of a tetrahedron.

A theory built on this basis directly is that of *tetrahedral shrinkage* (Figure 18.2). It surmises that our globe started as a hot body. Due to cooling, a solid skin (corresponding to today's crust) formed, which, upon further cooling, became too big for the shrinking interior. The shrinkage produced the form of a tetrahedron, simply because, of all regular bodies, it has minimum volume for a given surface. The corners of the tetrahedron would correspond to the continents; the faces to the ocean basins. The chief argument against the theory is that the crust of the Earth simply does not have rigidity properties which would allow it to stand up above a shrinking interior in such a manner. It is much more likely that the crust of the Earth would adjust itself under these conditions by forming folded mountains than by standing up

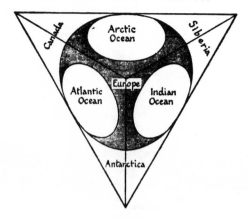

Figure 18.2 *The tetrahedral distribution of oceans and continents. Pacific ocean on concealed face (after A. Holmes,* Principles of Physical Geology)

to form continents. The tetrahedral shrinkage theory was conceived before it was known that the Earth's crust is thicker in continental than in oceanic areas, and therefore a mechanism was sought which would merely elevate parts of the ocean bed to form the continents.

Another theory which seeks to explain the tetrahedral distribution of continents (again assuming such a distribution as original), but which also tries to explain the fundamentally different structure of the continental crust as compared with the oceanic crust, is based upon the idea of convection currents. Forms of the mechanism have been proposed by many people including the Dutch geophysicist Professor Vening-Meinesz, Professor B. Gutenberg and Professor A. Holmes. It assumes that the interior of the Earth, at least in primeval times, was hot and liquid enough to permit convection currents like those in a pot of water heated from below. The heat flux required to cause the convection was due either to the process of cooling of the Earth, or else to the release of energy from decaying radioactive elements. In time, a differentiation of the original material of the Earth took place into lighter and heavier fractions, the heavier fractions sinking into the core and the lighter ones floating and collecting as "scum" in places where the convection currents descended. The "scum" would form the continents. Convection currents tend to arise in a cellular pattern, and it is possible to envisage an octahedral one

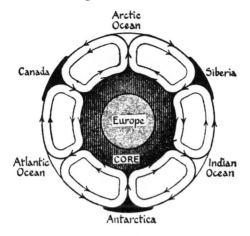

Figure 18.3 *A section through the Earth showing the sub-crustal convection currents that might give rise to the tetrahedral configuration*

with four descending, and four complementary ascending currents (Figure 18.3). If the descending currents gave rise to the continents, the latter would have collected at the corners of a tetrahedron.

The more general theory, that the continents were formed in some less specific fashion by convection currents, is today the most widely accepted one. It is not necessary to assume that the primeval convection currents were octahedral. One can suppose different patterns; a *single* primeval continent may have been the result of one such pattern. In this case the hypothetical continent is called Pangea. Alternatively, if *two* continents were formed at the beginning, they are generally referred to as Laurasia and Gondwanaland (Figure 18.4). Objections to the convection theory are that the physical properties of the Earth's mantle would not permit such mobile behaviour, and that currents of this kind would not produce the types of ocean coast, or patterns of mountain belts which exist.

A further theory of the origin of continents is the hypothesis of their formation by *expansion* (Figure 18.5). It assumes that the Earth was much smaller at the beginning than it is now, having a diameter of about one-half its present one. Somehow, a "crust" was formed on it, which was everywhere some 30 km thick. Then, as the diameter grew, the original crust broke up and its remnants

Figure 18.4 *The continental shields making up the two supercontinents of Laurasia and Gondwanaland (after A. L. du Toit,* Our Wandering Continents)

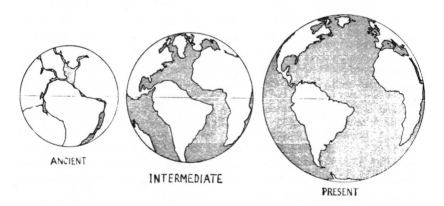

Figure 18.5 *How the Earth's oceans could have developed by global expansion (after O. C. Hilgenberg,* Vom Wachsenden Erbfall)

are the present continents. Expansion was supposed to have started by ocean "cracks" like those recently proved in the Atlantic Ocean, south of Africa and across the Indian Ocean. An increase by a factor 2 in diameter represents a surface increase by a factor 4, which produces about the right order of relative area occupied by the present-day continents. However, such a radius increase also produces a volume increase, and a corresponding density decrease, by a factor 8. Since the present average density of the Earth is about $5\frac{1}{2}$ times that of water, the average density before the start of the expansion would have been about 44 times that of water. It is difficult to explain this. Certainly simple heating could never produce such a large density difference. A corresponding physical change of the material inside the Earth also seems difficult to imagine. Advocating a change in the universal constant of gravitation appears equally far fetched.

The above discussion shows that only the hypothesis of the formation of continents by convection is left as reasonably possible although it, too, has many weak points. We do not know whether the Earth was hot and liquid at one time of its life and therefore whether the postulated convection was at all possible. A new theory, not yet conceived, may well provide the final answer to the question of how the continents were created. Many of our uncertainties will be resolved once we can be more definite about the nature of the Mohorovicic Discontinuity. After all, the theories of continent formation rest upon the assumption that the land-masses are fundamentally different from oceans. Should it turn out that the Mohorovicic Discontinuity represents a purely physical change rather than a chemical change, we shall be faced with two possible states of equilibrium in the continental and oceanic parts of the crust respectively. We must then seek the thermodynamic reasons why the material in one condition should concentrate in the continental regions, and in its other form in the oceanic areas. The problem of the nature of the Mohorovicic Discontinuity will certainly be solved once it becomes possible to drill a hole through it, and then scientists will be able to give a more definite answer to many questions concerning the Earth's evolution.

26 October, 1961

19

The Upper Mantle Project

GEORGES LACLAVÈRE

At the General Assembly of the International Union of Geodesy and Geophysics, starting in Berkeley, California, the world's geophysicists will review plans for a concerted effort in exploring the nature of the Earth at a depth of many miles.

The development of seismology over the past fifty years, both theoretical and instrumental, has permitted a detailed analysis of the propagation of seismic waves in the interior of the Earth. It has led to the conclusion that our planet consists of three regions, a central core of some 3500 km radius surrounded by a solid mantle nearly 3000 km thick on which lies the thin outer crust. Between these regions various discontinuities have been identified: when seismic waves cross the various layers their velocity changes, indicating a variation in the physical and chemical properties of the interior of the Earth.

The outer crust is thin. Its thickness decreases down to as little as 6 km under the deep ocean floor, its maximum reaches about 50 km, or perhaps more, under the highest mountains. The surface of separation between the crust and the mantle is known as the Mohorovicic Discontinuity after the name of its discoverer, a Yugoslav seismologist, in 1909. While the mean density of the crust, deduced from geological information and confirmed by geophysical methods, is about 2.7, the density of the mantle is much higher. It has been estimated to be 3.3 in the upper layers, increasing to 5.5 at the core boundary.

There is evidence that the mantle is the seat of dynamic processes. Deep earthquake foci have been identified at depths of several hundred kilometres. The theory of convection currents created by thermal gradients, horizontal and vertical, suggested by Vening Meinesz, seems to be the more widely accepted theory for the generation of dynamic processes in the mantle. It is logical to believe that movements in the mantle will influence deformations in the thin and light crust. These deformations cause earthquakes, vulcanism, mountain-building and other surface changes. They may result in the displacement of entire continents moving slowly as rafts on the discontinuity layer, if the theory of continental drift is accepted. Action in the mantle may also generate metallogenic processes bringing concentrations of useful minerals into the crust.

Unfortunately, in spite of extensive research carried out for many years past by geophysicists, our knowledge of the Earth's interior is still very poor. Direct approaches limited to geological evidence cannot penetrate more than a few kilometres into the Earth and, in spite of elaborate techniques developed in the United States and in the USSR, the "Moholes" have not yet reached to the bottom of the crust and samples of the mantle material have not yet been collected.

Recognising the need for a better knowledge of our planet's interior and the difficulty of obtaining it, the International Union of Geodesy and Geophysics, at its General Assembly in Helsinki in 1960, proposed, at the instigation of Professor V. V. Beloussov, to undertake a concerted international attack on the problems involved. This is the Upper Mantle Project. A committee has been formed and the cooperation of other International Scientific Unions has been invited: i.e., Geology, Astronomy, Physics, Chemistry, Mechanics and Crystallography. An appeal to collaborate was sent out to national committees, to interested scientists and to relevant organisations. The response is very encouraging and a number of countries have already prepared programmes of investigation.

The objectives of the Upper Mantle Project are to determine the variation of the physical and chemical properties of the upper mantle and crust, both radially and with latitude and longitude, the present and past movements of material in the mantle and the sources of energy within it. These studies are a necessary prerequisite to the understanding of the causes of tectonic (moving rock) and magmatic (molten rock) processes.

To this end, development and application of new geophysical techniques will be fostered, including magnetotelluric measurements of electric currents in the Earth, studies of long waves and free oscillations of the Earth, deep seismic soundings, ocean bottom seismographs, deep drilling, heat flow determinations and studies of rock magnetism. Along with observations by these methods on the present state of the mantle, data will also be considered from geological records in the various continents on the evolution of tectonic and magmatic processes through the Earth's history. Conventional geophysical methods (geodetic, gravimetric, seismic, etc.) will be used concurrently.

For the implementation of the project, the Earth has tentatively been divided into several main tectonic zones. On the continents, the following zones have been suggested: shields and stable zones, folded and activated zones and continental margins. For the oceans the division are: deep ocean basins, submarine ridges, inland seas and island arcs.

It is obvious that the collection of samples of the upper mantle in different zones would provide most valuable information; but, until rapid deep drilling machines have been developed, it seems difficult to believe that many "Moholes" can be drilled. If they are drilled through the ocean floor, where the distance through the crust is smaller, apart from the difficulty of operating the drilling equipment on board a ship (which seems to have been overcome), the basalt layer that constitutes the ocean floor may be an impenetrable obstacle. On the other hand, on the continents, the distance to be drilled is much longer. In order to overcome these difficulties, Goguel, in France, has suggested that upper mantle material could be obtained from inclusions in the Kimberlite volcanic pipes. Beals, in Canada, offers another original solution from the study of meteorite craters. There are large circular features, identified from air photographs, in particular in Hudson Bay and in the Gulf of St Lawrence, which are several hundreds of kilometres across, if these were caused by meteoric impacts the crust of the Earth should be partly missing and this would allow the mantle material to rise above its usual level and to become accessible to the drill.

It is hoped that the coordinated international approach to the study of the upper mantle will give the answers to some unsolved problems, such as: (1) differences in the upper mantle and the crust as between continents and oceans, and as between the various tectonic zones; (2) mechanism at the focus of an earthquake; (3) the nature of the Mohorovicic Discontinuity; (4) variations of electric conductivity with depth; (5) the origin of the "low velocity" zones in the upper mantle; (6) the extent and significance of intermediate layers; (7) evolution processes deep in the Earth during geological time; and so on.

A major question still to be elucidated, the solution of which may come out of the Upper Mantle Project, is: Are the continents drifting or not? In 1912 the German meteorologist Alfred Wegener made the suggestion that, if land could move slowly up and down, it could as well move sideways. He produced evidence from geological, palaeoclimatic and palaeontological observations that the continents were slowly drifting at the rate of a centimetre or so a year and he suggested that they originated from a unique continent, "Gondwanaland", which was dislocated a few million years ago. He explained thereby the similarity of the delineation of the coasts of Africa and South America on the two sides of the Atlantic Ocean. Vigorous controversy took place in the 1920s about his theory, one major obstacle to its acceptance being that it was not possible to justify the force required to make the continents drift, and that the accuracy of geodetic measurements was not sufficient to detect such small displacements.

The theory, which was abandoned for several decades, is gaining new support but is far from being widely accepted. Opponents declare that the continents have been growing through the geological ages, by the addition of peripheral ranges, and they accept the concept of a rigid Earth with permanent continents and ocean basins. Mountain-building processes are explained by some as a consequence of contraction from cooling. But it seems from recent studies that radioactive processes at work within the Earth produce more heat than can be dissipated by radiation and here Vening Meinesz's theory of convection currents explains mountain formation in a non-shrinking Earth.

A recent theory about the deformation of the crust takes into account a proposal advanced by Dirac that the gravitational constant is slowly decreasing with absolute time. The consequence would be to release the pressure inside the Earth and cause a slow expansion of its radius.

This theory would give a satisfactory explanation of the system of mid-ocean ridges revealed in 1956 by M. Ewing and B. C. Heezen. They suggested that the numerous submarine chains of mountains, which hydrographic soundings have spotted here and there, form a single system, consisting of the mid-Atlantic ridge which turns south of Africa, across the Indian Ocean and south of Australia, then across the Pacific Ocean to the Mexican Coast. This remarkable submarine mountain ridge extends for nearly 40 000 km; it is hundreds of kilometres wide and rises several kilometres above the ocean floor; it is dotted with active volcanoes and its rocks are unusually hot, indicating a very high heat flow. A slow expansion of the crust throughout the geological ages, causing a crack where it is thinner, would explain the formation of the mid-ocean ridge and the unusual heat flow there by the exposure of deeper layers.

Among such a variety of contradictory theories the scientists have to make a difficult choice. A symposium on the upper mantle being organised in Berkeley, California, will be the core of the 13th General Assembly of the International Union of Geodesy and Geophysics later this month. It will give an opportunity for reports to be presented on the upper mantle studies which have been initiated since the Helsinki General Assembly in 1960. This symposium, which will last five days, will touch on many of the subjects mentioned above and will, no doubt, give a great impetus to this new international venture.

15 August, 1963

20

Taking continental drift seriously

PETER STUBBS

After nearly 60 years of often heated dispute, the theory of continental drift seems to have come to stay. Geophysical thinking along many lines is converging in support of the theory.

That satellite measurements can have anything whatsoever to do with theories of continental drift may seem a remote possibility. Yet it is on the interpretation of data obtained in this way that one of the crucial arguments about the nature of the Earth's interior hangs, and hence a large part of the discussion as to whether or not it is physically feasible that the landmasses of the Earth should formerly have changed their positions – or are still in the process of moving. More than 50 years after Alfred Wegener launched his theory on a sceptical world, it is interesting to see that the theory of continental drift is losing none of its appeal as a subject for debate. Indeed, nowadays it provides more than ever the strong stimulus behind an increasing number of new lines of research in some very diverse fields of study. Progress in many of the most exciting of these provided the topics for a two-day intensive meeting held last week at the Royal Institution in London, under the auspices of the Royal Society, such recognition being itself a clear indication of the respectable status which the theory has now acquired.

The early workers, including Wegener, du Toit and Taylor, had to rely, of course, on purely geological evidence to support their theories as to how the various continents, like Africa and South America, were once joined together. They used such things as fossil stratigraphical and structural similarities, the evidence of former glaciations and other indications of past climates. In general, most of the detailed theories postulated that there were formerly two major super-continents, termed Laurasia in the

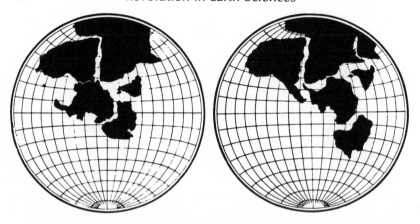

Figure 20.1 *The southern hemisphere continents reconstructed into Gondwanaland according to (Left) Professor J. T. Wilson and (Right) A. L. du Toit (after K. M. Creer)*

northern hemisphere and Gondwanaland (Figure 20.1) in the southern. At some point in geological history these split up into the continents more or less as we know them. Two predominant movements were alleged to have led to the present distribution of the land masses; a radial movement of the Gondwana continents away from the South Pole, and westerly drift of the Americas away from Africa and Europe.

The evidence on which all this was based, however, was controversial and inconclusive. Moreover, many geophysicists refused to accept the idea, saying that the Earth's mantle underlying the crustal layer was far too strong to permit such geographical caprices. The turning point in the attitude of the scientific world to this apparently dubious theory came just over a decade ago when studies of the fossil magnetism in rocks led to an entirely unexpected and completely objective method of examining continental drift. Since then the results which the palaeomagnetists have produced have clearly had their effect upon geophysicists generally. At the Royal Society meeting, palaeomagnetism, oceanography and theoretical geophysics predominated, to the exclusion of the stratigraphy and palaeontology which once prevailed at similar discussions.

Briefly, the principle underlying palaeomagnetic research is that by making measurements of the directions of "fossil" magnetism preserved since formation, in samples of rocks of known age one can calculate the past latitudes and orientations of each of the world's landmasses (see "Rock magnetism and the movement of continents", the first item in this section). The picture is complicated by the possibility of polar wandering – the movement of the Earth's crust as a rigid whole relative to the rotational poles, as opposed to relative movements between continents – but criteria have been developed for telling the two apart. Taken with tests of the magnetic stability of the rocks, the several hundred sets of results obtained for geological formations all over the world now leave little room for doubt that continental drift certainly occurred. It would need some very curious artefact to account for the results in any other way, and most of the data fit in tolerably well with the general pattern envisaged by Wegener.

Dr K. M. Creer, of the University of Newcastle upon Tyne, has recently managed to produce palaeomagnetic latitudes for a series of formations in South America, the last of the major continents for which such information was lacking. Interpreted in their simplest fashion, they show that South America probably drifted across the South Pole since Cambrian time 600 million years ago, as shown in Figure 20.2. Adding his own results to those of people working on rocks from the other major continents, Dr Creer has been able to test the reconstructions of Gondwanaland suggested

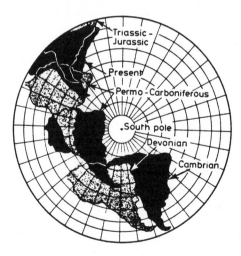

Figure 20.2 *South America appears to have drifted across the South Polar region, according to Dr Creer*

by du Toit, and more recently by Professor J. Tuzo Wilson (Figure 20.1), of Toronto University, and to make an attempt to decide when it began to break up and drift apart.

The palaeomagnetic data do not conflict with either of the two reconstructions, though they are not quite adequate to distinguish which is the more plausible. By plotting the rate of change of latitude with geological time for each of the major continents, Dr Creer showed that Africa, Australia and South America, at least, all underwent a large change of latitude during the Palaeozoic Era, approximately between 600 and 250 million years ago. During the subsequent Mesozoic Era there appeared to have been a similar, though smaller, change of latitude of Africa, India and Europe. He thinks that Gondwanaland could certainly have existed during the Palaeozoic Era but not later if the magnetic record of rocks is correct. The change of latitude in the Palaeozoic Era he attributes to polar wandering which he believes to be the more important geophysical process, continental drift merely being a minor phenomenon incidental to more profound processes going on inside the Earth.

Thus it seems likely that the breakup and drift apart of the continents may have begun some time in the Mesozoic Era, roughly 250 million years ago; Africa and South America probably reached positions near their present ones early in the drift process, and the other continents arrived later.

Professor S. K. Runcorn of the University of Newcastle upon Tyne, one of the most frequently heard exponents of modern continental drift theory, showed several years ago that the polar wandering paths for Europe and North America deduced from palaeomagnetic measurements were displaced in a manner suggesting the drift apart of the two land-masses. Similar recent work by Soviet workers, carried out on rocks from either side of the Urals, shows that the two halves of Russia have hinged about a point on this line of geological crumpling. Another interesting development in the Soviet Union has been the other half of the Indian continental drift story. India, as Dr J. A. Clegg and his colleagues at Imperial College, London, showed, seems to have drifted from a point well down in the southern hemisphere comparatively late in geological time; the fact is compatible with the folding of the Himalayas by the approach of India to Asia. But the point at issue was whether or not this was true relative movement between the two land-masses, or whether Asia had perhaps drifted northwards at the same time. The new results from Russia confirm that these two land-masses seem to have shortened the original distance between them by about 3000 miles (4830 km).

Geographical correspondences between the shapes of coastlines, notably that between the east coast of South America and the west coast of Africa, were one of the earliest features that led scientists to think of continental drift. South America could be tucked into Africa very conveniently. Rationalising this idea, which has previously been little more than a subjective impression, Sir Edward Bullard and a team from the Department of Geodesy and Geophysics, Cambridge, are currently in the middle of a global analytical operation in which they have persuaded a computer to make the best possible fits between pieces of continent that may once have been adjacent. The researchers are using recent American hydrographic maps for the purpose. The computer works out the point on the globe about which it is necessary to rotate a given land-mass, and the angle through it must be turned, to get the best fit; it then computes the error or misfit.

The picture which emerges shows how remarkably well the northern hemisphere continents socket into one another at the 500-fathom line. The Cambridge workers have a similar correspondence for South America and Africa. Though there are some holes unaccounted for they are not unexpected and the hole in the North Atlantic, for instance, could well be filled by the island of Rockall which is possibly a continental fragment. However, it is remarkable that the coastlines of the continents should not have suffered a larger degree of erosion or distortion due to the deposition of new rocks and, if they have been torn apart, that there should not have been a rather large amount of stretching. There was some scepticism at the meeting that these good fits – with mean errors of less than one degree – should really represent the original outlines of the bits of the continental jigsaw puzzle.

The fit receives support from the work of Dr J. A. Miller, of Cambridge, who has plotted on a regional scale the distribution of rock-formation ages, determined by radioisotopic methods, over the map reconstructed in this way. He finds that they lie on concentric zones with the younger rocks towards the outside and the older ones round Hudson Bay; there also seems to be another centre of older rocks towards northern Russia. More work is necessary to fill in this picture but if it is valid it would seem to bear out the continental reconstruction. Eventually it will be necessary to test Sir Edward Bullard's detailed results against accurate palaeomagnetic data.

If the oceans are the rifts left by the tearing apart of continents, one would expect them to be very different in their physical characteristics from the continental parts of the Earth's crust. Some of the world's best oceanographers, present at the meeting, demonstrated not only that this was abundantly so but also that workers in the subject are increasingly seeking to explain their discoveries in relation to continental drift. The developing picture is evidently much more complex than early oceanographic work led people to suppose.

Professor Tuzo Wilson, for instance, put forward an attractive theory based on the assumption that the oceanic ridges (see Figure 20.3), originally considered to form a single system running through the centres of the major oceans, are the lines along which slow sub-crustal thermal convection currents rise from the depth of the Earth's mantle. The idea of sub-crustal convection currents, which was proposed in the first place to explain the forces that

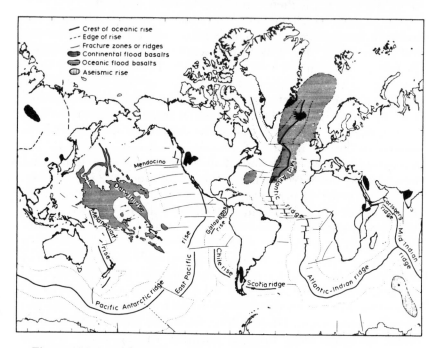

Figure 20.3 *The major oceanographic features discovered up to the present, showing ocean ridges, numerous tear-faults and the great lava outpourings of the north Atlantic and ancient Darwin Rise (Professor H. W. Menard)*

squeeze up mountain chains, is also the mechanism that has been most favoured to bring about the continental drift. Professor Wilson's idea is that the rising columns of material beneath the oceanic ridges spread out on either side and flow towards the continental borders to sink once more beneath the mountain chains. He claims that such a process would account for volcanic activity and increased thermal flow along the ocean ridges and that there is evidence, particularly in the Hawaiian Islands, of flow lines that actually show the mechanism at work. Older volcanoes, he says, are to be found farther out from the ridges than the younger ones, having been carried there on the outward-flowing "conveyor belt".

While this explanation may fit what is known about the structure of the Atlantic Ocean, the latest information on both the Pacific and Indian Oceans shows that their structures are more complicated and that their ridges are neither central nor behave in so straightforward a manner (Figure 20.3). Professor H. W. Menard, of Scripps Institute of Oceanography, La Jolla, California, pointed out that recent exploration in the Pacific had shown that the East Pacific Rise is supplemented by a least two other ridges, the Chile Rise and the Melanesian Rise. Neither the areas of high heat flow nor the distribution of volcanoes corresponded, in general, with the positions of any of these ridges, while the rift characteristic of the centre of the Mid-Atlantic Ridge, in the case of the East Pacific Rise was frequently displaced 500 to 1000 km from the crest.

It is also apparent, says Professor Menard, that ocean ridges can both rise and fall. By measuring the depths of the wave-cut tops of former mid-Pacific volcanic islands, now represented by coral atolls, one can show that they once lay along an ancient ocean ridge, the Darwin Rise, which has since sunk back into the ocean floor. Thus the oceanographic evidence does not fit itself into the continental drift theory in any simple way, though it may not contradict the hypothesis and, in fact, suggests in certain parts of the globe how the process takes place; as Dr R. W. Girdler, of Newcastle, demonstrated, the physical features of the Red Sea Rift imply that here an incipient ocean is actually being created. The Gulf of California may well be another.

That something very like Professor Tuzo Wilson's mechanism is happening along the Mid-Atlantic Ridge, however, is borne out by some remarkable research on the volcanic rocks of Iceland. It is the biggest island athwart the Ridge, and almost entirely com-

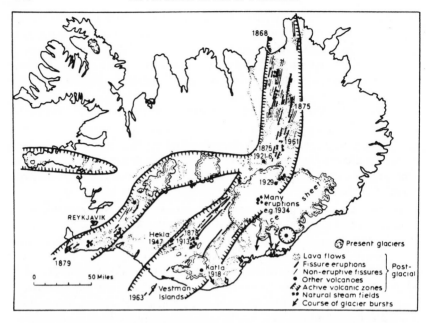

Figure 20.4   *Iceland, situated astride the Mid-Atlantic Ridge seems to be splitting apart. This map, due to Dr G. P. L. Walker, shows the most recent volcanicity is confined to the island's central region where lavas are extruded through long fissures. Dates indicate eruptions, including last year's new volcano off the south coast*

posed of lava flows which are fed with molten material from fissures rather than central vents. Dr G. P. L. Walker, of Imperial College, London, believes that Iceland is gradually being torn apart. The younger lavas lie in the island's centre (Figure 20.4), and it looks as though the vertical "feeders" supplying the volcanoes are the result of cracks due to the stretching of the island. Dr Walker has made an estimate of the stretching from the known thickness of the lava flows, and the number and thickness of the feeders that would be needed to supply them. The total lava pile below Iceland is probably about 20 km thick and would need feeders totalling some 400 km of cracks. If this is so, Iceland has been dilating at roughly one cm per year over the latter part of its geological history, and is still expanding. This is a highly realistic manifestation of continental drift.

As Figure 20.3 shows, seismic evidence has established the presence of an enormous outpouring of lava over the northern third of the Mid-Atlantic Ridge. The former Darwin Rise seems to be a similar phenomenon. Professor Menard estimates that these lava sheets may contain of the order of ten million cu. km of basalt – far greater than anything like them which occurs on the continents. They would conveniently form the necessary sources of material to replace continents that have drifted apart.

However, the facts of physics must be satisfied, and this is where the satellite measurements come in. Accurate tracking of their orbits gives an excellent method of finding how the gravitational field of the Earth varies over its surface and hence of giving a detailed picture of its exact shape. Results prove that the Earth's equatorial bulge is too big to be in equilibrium at the planet's present rate of rotation. From what we know about how the Earth has been slowing down, the bulge corresponds to the rate of rotation about ten million years ago. This means that the mantle of the Earth has to support the extra bit of the bulge against gravity. Dr G. J. F. MacDonald argues that the mantle must be so strong to be able to do this that convection currents would never be able to occur in it.

Further analysis of the Earth's shape indicates that "highs" occur over certain regions and "lows" over others. Professor Runcorn believes that the highs are, in fact, above the culminations – sort of slow fountains – of sub-crustal convection currents. From the shape maps he has constructed, on this basis, another map of the crustal stress patterns that would be associated with the highs and lows. It certainly agrees rather well with most of the ideas about continental drift, but the topic was one of considerable dispute at the meeting.

Whatever the ultimate conclusion about the satellite data, one piece of evidence for continental drift is irrefutable. As Professor V. Vacquier, of Scripps Institute, emphasised, one of the significant things to come out of recent oceanography is the prevalence of trancurrent or tear-faults on the ocean floors (see Figure 20.3). These are large relative movements of adjacent pieces of the Earth's crust that slide past each other in a *horizontal* plane. Their amounts of movement can be gauged directly by the way that patterns of magnetic anomalies over the floor are displaced. Many of them indisputably represent translocations of hundreds of miles – the Mendocino Fault, for example, consists of a shift of some

700 miles (1100 km). On land, too, more and more faults of this type are being discovered. The Atacama Fault along the coast of Chile is a tear of some 500 miles (800 km), responsible for the disastrous Santiago earthquake of May 1960. The still active San Andreas Fault of California is a well-known tear of similar size. As Professor C. R. Allen of California Institute of Technology, remarked, here is continental drift taking place before our eyes at a considerable rate.

26 March, 1964

21

New light on the Earth's interior

A. E. RINGWOOD

Earth scientists have argued for a long time whether the layers of the Earth's interior differ from one another chemically or merely physically. Soviet engineers are just starting to drill the Earth's crust to probe part of the problem directly; but new laboratory experiments described here have already begun to provide the answer.

A primary objective of the Earth sciences is to attain a comprehensive understanding of the present physical and chemical constitution of the Earth's interior. This objective is important in its own right. It is also an essential prerequisite to the exploration of other fundamental problems, such as the origin of the Earth, its subsequent evolution, and of the processes in its interior which are responsible for such phenomena as mountain building, volcanism and earthquakes. During the last 10 years, the application of experimental high pressure and temperature techniques to the study of the Earth's interior has resulted in some major advances in our knowledge of constitution.

Our principal source of physical information on the Earth's interior comes from seismology – in particular, from the way in which seismic compression (P) and shear (S) wave velocities vary with depth (Figure 21.1). These divide the Earth into three principal regions – crust, mantle and core. The crust is about 40 km thick in continental regions where it is highly heterogeneous, composed largely of rocks varying in composition between granite and basalt. In deep oceanic regions, the crust is about 5 km thick on the average and is believed to be dominantly of basaltic composition. The seismic P-wave velocity of the crust varies between 6 and 7 km/s. Below the crust, lies the mantle, which extends downwards for 2900 km. In most regions the crust is separated from the mantle by the "Mohorovicic Discontinuity" or "moho". This is a narrow zone over which the seismic P velocity increases suddenly to about 8.2 km/s. There has been a great deal of discussion in recent years about the nature of the moho and whether it is caused by a change in chemical composition or by a change in physical state – that is, a phase change. One of the aims

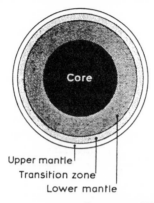

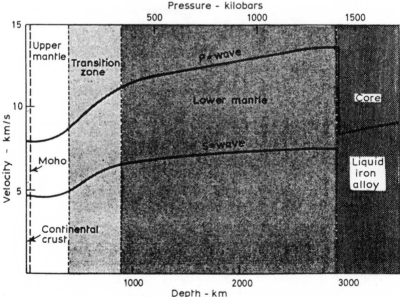

Figure 21.1 *Principal subdivisions of the mantle based upon the variations of seismic P- and S-wave velocities (1 kilobar is approximately equal to 1000 atmospheres pressure)*

of the ill-fated Mohole project was to resolve this issue. Later in this article I shall describe some laboratory experiments which have recently provided decisive evidence on this point.

The mantle transmits both P and S waves and hence is solid, not liquid. From Figure 21.1 we see that the mantle can be divided into three regions on the basis of the variation of seismic velocities with depth: the *upper mantle*, extending down to about 400 km is characterised by generally low P and S velocity gradients, and by substantial regional variations in seismic velocities; the *transition zone*, extending between about 400 and 900 km, is characterised by a very rapid increase of seismic velocities with depth; and finally the *lower mantle*, from 900 km to 2900 km, is character-ised by a moderate and uniform rate of increase of seismic velocities. One of the principal achievements of experimental investigations during recent years has been to demonstrate the nature and cause of this threefold subdivision of the mantle. I will be returning to this topic later on in this article.

The nature of the moho

It has long been known that the dark-coloured or "basic" rocks which are rich in iron and magnesium may crystallise in two forms: as *basalt*, composed principally of the minerals plagioclase and pyroxene; and as *eclogite*, composed of the minerals garnet and omphacite. Although these rocks have approximately the same chemical composition, their physical properties are very different. Basalt has a density of 3.0 g per cu. cm and seismic P velocity of about 6.8 km/s, whereas eclogite has a density of 3.5 g per cu. cm and a seismic velocity of about 8.3 km/s, which is similar to the velocity in the mantle immediately beneath the crust. It was realised by Sir Lewis Fermour as early as 1912 that eclogite must be a high pressure form of basalt, and he went on to suggest that the upper mantle was composed of eclogite. This proposal was developed further by Arthur Holmes. He pointed out that since the lower crust was probably composed of basaltic rocks while the physical properties of the upper mantle were identical with eclogite, the moho might represent a phase change between basalt and eclogite. In the period 1930 to 1950 this hypothesis became somewhat neglected. Geologists favoured an alternative view that the upper mantle was composed of ultrabasic rocks such as peridotite, which is dominantly composed of the mineral olivine. This rock also possessed a seismic velocity – identical with that of eclogite – and density which matched the properties of the upper mantle. Accordingly, on this hypothesis, the moho would be caused by a chemical change from intermediate-to-basic lower crust, into ultrabasic upper mantle.

More recently, the phase-change hypothesis has been revived, and has won support in many quarters. An attractive property is that changes in temperature at the moho might transform basalt to eclogite or vice versa, with large accompanying volume changes. These might cause the uplift and depression of the crust inferred by geologists but very difficult otherwise to explain.

During the past 10 years the development of experimental apparatus capable of duplicating the range of pressures and temperatures occurring in the upper mantle has meant that the phase-change hypothesis can now be investigated by direct experiment. It is clearly of decisive importance to establish the pressures and temperatures at which the transition occurs, the detailed mineralogical equilibria involved in the transition, and the way in which these control the change in physical properties caused by the transition. Preliminary investigations by workers at the Geophysical Laboratory of the Carnegie Institution of Washington succeeded in outlining, within rather broad limits, the conditions over which basalt and eclogite are stable. However, the experiments were not sufficiently detailed to permit an evaluation of the phase-change hypothesis.

Recently, at the Australian National University, Dr David Green and I have completed a detailed investigation of this transition during which we made more than 200 experimental runs at high pressure and temperature, and investigated the resulting mineral assemblages in detail by petrographic, X-ray and electron-microprobe techniques. We have succeeded in specifying the pressure and temperatures required for the transformation of a series of naturally occurring basalts into eclogites and the nature and width of the transformation interval.

Some of the results are set out in Figure 21.2. Notice that the transition from basalt to eclogite is not sharp; instead there is a wide transitional interval (garnet granulite) composed of the minerals garnet, pyroxene and plagioclase, which intervenes between the basalt and eclogite fields. We were able to show that, to a first approximation, the changes in seismic velocity and density associated with the transition occur uniformly over this transition interval, which ranges from 3 to 12 kilobars in width,

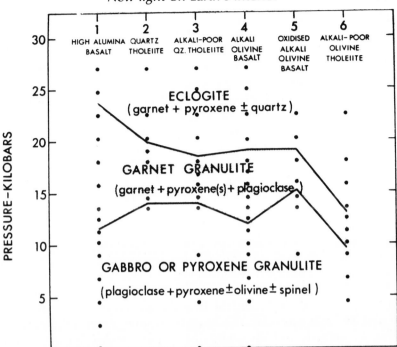

Figure 21.2 *Diagram showing the results of experiments on several typical basalts at 1100°C over a range of pressures. Each point represents a single experiment. In each case, the low-pressure basaltic or gabbroic mineral assemblage passes gradually, over a broad pressure interval, into the high-pressure eclogitic mineral assemblage (garnet granulite)*

equivalent to a depth interval of 10 to 40 km. Accordingly, it is impossible to obtain a sharp increase of seismic velocity from 7 to 8 km/s in a small depth interval of 5 km or less as is required if the moho is to be explained by the phase-change hypothesis.

This difficulty is compounded by the tendency of temperature-depth curves in the Earth to intersect the transition-field boundaries at high angles thus further broadening the effective transition width. Notice also the wide range in pressures required to stabilise eclogite derived from several basalts which differ in chemical composition by rather small amounts. This property would make it difficult to understand the relative uniformity of continental thickness in stable regions of the Earth's crust. Finally, we were able to show that the temperature gradient of the basalt–eclogite transition was inconsistent with these transitions occurring generally under the pressure and temperature conditions existing at the moho – except, perhaps, in limited regions. Thus, a combination of arguments based directly upon laboratory investigation, make the phase-change hypothesis of the moho extremely improbable and demonstrate that the moho is most likely caused by a change in *chemical* composition from basic lower crust into ultrabasic or peridotitic upper mantle.

The transition zone in the mantle

The physical and chemical properties of the upper mantle are consistent with a mixture of rocks composed of varying proportions of the common minerals olivine, pyroxene and garnet, in the stated order of abundance. However, at a depth of about 400 km, seismic velocities begin to increase rapidly (Figure 21.1) and soon become too high to be caused by these minerals. It was demonstrated by Professor K. E. Bullen in 1936, that the rapid increase in seismic velocity which characterises the transition zone was accompanied by a corresponding rapid increase in density. Many hypotheses to account for the nature of this region have since been proposed, one of the most popular being that it was caused by an increase in the amount of iron in the mantle – perhaps the iron occurred in the native state as in some meteorites. Professor J. D. Bernal, however, together with Sir Harold Jeffreys, proposed a different hypothesis. They suggested that olivine, which is believed to be the most abundant mineral in the upper mantle, may transform itself into a different crystalline structure – the spinel structure – at a depth of about 400 km (equivalent to a pressure of 130 kilobars), accompanied by marked increases of seismic velocity and density.

This hypothesis was generalised and expanded by Professor Francis Birch at Harvard, in the course of a classic investigation of the elastic properties of the mantle. He suggested that in the transition zone a series of major phase transformations took place according to which the common upper-mantle minerals olivine, pyroxene and garnet were transformed into new, close-packed crystalline forms, possessing much greater densities and higher seismic velocities. The lower mantle was interpreted as being composed of these close-packed minerals, in which further transformations under increased pressures were not possible. This region was essentially homogeneous, and characterised by only a moderate increase of seismic velocity with depth.

Decisive evidence in favour of this hypothesis has been obtained during the past 10 years. The pressures required for these transitions were very high – between 130 and 300 kilobars – and, until recently beyond the range of static high-pressure – temperature apparatus. The evidence was thus first gathered by indirect

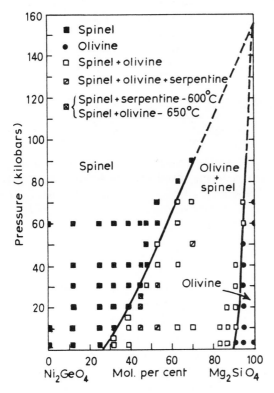

Figure 21.3 *Diagram showing the phases present in the system Ni$_2$Ge$_4$ (spinel)-Mg$_2$SiO$_4$ (olivine) at 600°C and 90 kilobars. Each square denotes the results of a single experiment. It is seen that the amount Mg$_2$SiO$_4$ which is able to enter the spinel solid solution increases regularly from 27 per cent at atmospheric pressure to 70 per cent at 90 kilobars. Extrapolation of the phase boundary indicates that pure Mg$_2$SiO$_4$ should be transformed to a spinel structure at about 155 kilobars*

methods, making use of the techniques of thermodynamics, crystal chemistry and by the extrapolation of phase boundaries determined at lower pressures (e.g., Figure 21.3). In this way, in 1956, I calculated on thermodynamic grounds that common olivine should change to a spinel structure at a depth of about 500 km. Soon afterwards, I discoverd the occurrence of direct olivine–spinel transitions in the compounds Fe_2SiO_4, Ni_2SiO_4 and Co_2SiO_4 at pressures between 20 and 70 kilobars. This was followed by work on solid-solution phase boundaries as in Figure 21.3 over a wide range of pressures. Using mineral analogues containing germanium in place of silicon, I showed that pyroxenes as a group were unstable at high pressures, changing into denser phases, including a structure first suggested by Birch – that of ilmenite.

A major discovery was made in 1961 by Stishov and Popova in Moscow, who showed that common quartz (density 2.65 g per cu. cm) could be transferred into a new phase with a density of 4.3 g per cu. cm at about 120 kilobars. Although this rutile-structure phase may not be a major constituent of the mantle, the transition is important; it demonstrates that at high pressures, the silicon atom present in nearly all rock-forming minerals, may change its coordination with respect to oxygen from fourfold to sixfold, thus permitting much closer packing of atoms and causing a very large increase in density.

The concluding stages of this investigation have now been reached. Last year, at the Australian National University, we constructed an apparatus capable of developing over 200 kilobars simultaneously with temperatures of about 900° C. We have been able to transform common olivine directly into a spinel structure accompanied by a density increase of about 10 per cent. The same apparatus has been employed to study the stability of common silicate pyroxenes and these, too, have been found to change into mixtures of spinel and rutile phases, or into new phases characterised by sixfold coordination of silicon.

It is now possible to state that the Bernal–Jeffreys–Birch hypothesis of the constitution of the mantle has been proven beyond reasonable doubt. Furthermore, the physical properties of the lower mantle when corrected to low-pressure–temperature conditions are found to be in close agreement with the properties of the new dense silicates which have been synthesised. Thus we now possess a broad understanding of the large-scale properties of the Earth's mantle in terms of the stability field and physical properties of the minerals present.

From the two examples discussed, we see that reasonably definite answers have been given to major, and formerly controversial, problems of the Earth's interior by the application of laboratory experimentation. This is only the beginning of the road. I expect that during the next 10 years, many of the problems which have perplexed geologists and geophysicists since the beginning of these sciences will yield to direct investigations using high-pressure–temperature techniques. It is difficult not to be an optimist in this field today.

16 March, 1967

PARASITES SUPPORT CONTINENTAL DRIFT

New support for the view that South America and Africa were connected in former times, as assumed by the continental drift theory, comes from parasitology. The fish parasite *Nesolecithus africanus,* discovered recently by Dr J. Dönges of Tübingen, belongs to the family of Amphilinidea and lives in the intestines of *Gymnarchus niloticus* (the "electric pike"). But while the host is found in all North African rivers, the parasite so far was found only in fish living in the River Yewa in Nigeria.

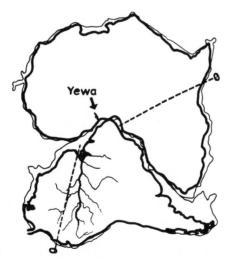

Figure 21.4

The nearest relative of *N. africanus* is *N. janickii,* found in the intestines of *Arapaima gigas,* the biggest South American fresh water fish populating the Amazon. Although the two parasites are different species they are very similar and must be assumed to have common ancestors in the not too distant past. Both of them only live in fresh water fish. Both of them use amphipoda living only in fresh water as intermediate host. According to Dönges it seems very unlikely indeed that some of their common ancestors crossed the thousands of miles of present day Atlantic.

If, however, South America and Africa were connected up to the Jurassic or Cretaceous times, as may be assumed from geological similarities, the rivers Yewa and Amazon were once near each other (see Figure 21.4). The common ancestors of the two Nesolecithus species then lived originally in one closed area which was later divided by continental drift (*Umschau,* vol. 17, 1967, p. 564).

"MONITOR", 28 September, 1967

DRIFTING ANTARCTIC YIELDS SPECTACULAR FOSSIL BED

On their first day of work in the summer antarctic season a team of nine American scientists have discovered a bed of fossil bones of several types of vertebrates, including amphibians and reptiles – the first such find in the Antarctic and one of the truly great fossil finds of all time. All the fossils appear to be remnants of now-extinct creatures that lived during the Triassic period – more than 200 million years ago. The discovery adds considerable weight to the theory of continental drift.

The finding includes part of a reptilian skull identified by Dr Edwin H. Colbert of the American Museum of Natural History in New York as *Lystrosaurus,* a reptile two to four feet long. "This is the key index fossil of the Lower Triassic period in the major southern land masses and establishes beyond further question the former existence of the great southern continent of Gondwana-land," said Dr Laurence M. Gould, internationally known geologist and board member of the US National Science Foundation, the chief sponsoring organisation for American scientists working in the Antarctic.

Dr Colbert described *Lystrosaurus* as a reptile "with a peculiarly shaped skull, the nostrils high on the skull between the elevated eyes. This," he said, "almost surely indicates aquatic habits." Fossils of these reptiles have been found in Asia and South Africa. Comparing the latest find with these specimens should provide further evidence that the Antarctic continent was once joined to other continents.

The fossils were found on 23 November, the first day of work in the field, said Dr David H. Elliot of Ohio State University's Institute of Polar Studies. They were discovered in a sandstone bed at Coalsack Bluff, in the Central Transantarctic Mountains about 400 miles from the South Pole.

Among other fossils discovered were bones of an extinct reptile, the thecodont. Thecodonts were ancestors of the dinosaurs, and fossil remains of these have been found in North America and Europe. Their evolutionary descendants are crocodiles and alligators and, through a more complex evolution, birds.

Another fossil remnant found by the Antarctic team was of an extinct amphibian called labyrinthodont. Two years ago scientists working in Antarctica found the first fossil from a labyrinthodont – the only other specimen known.

Discussing the find, the Ohio scientists noted that it is of great significance to students of Earth history. During recent years the continental drift theory has received increasingly favourable attention from geologists. "This theory," they pointed out, "developed in detail more than fifty years ago, supposes that the continents today are remnants of a once supercontinent, or maybe two such large continents, that broke up, the separate sections then drifting slowly across the face of the globe to their present positions. If this theory is valid then Antarctica was once a part of a great southern land mass known as Gondwanaland.

"The presence of fresh-water amphibians and land-living reptiles in Antarctica, some 200 million years ago, is very strong evidence of the probability of continental drift because these amphibians and reptiles, closely related to vertebrate animals of the same age on other continents, could not have migrated between continental areas across oceanic barriers."

"MONITOR", 11 December, 1969

FOSSILS FIT THE STORY OF CONTINENTAL DRIFT

Geologists believe, largely on the basis of palaeomagnetic data, that all of the present continents were grouped in one land mass, Pangaea, before about 800 million years ago, and that this then broke up and drifted first into two super-continents and then, gradually, into the pattern we see today. Now, studies of the fossil remains of early shellfish have revealed a distribution pattern very much in line with the palaeomagnetic evidence.

Mark McMenamin, of the University of California at Santa Barbara, has been looking at the distribution of the so-called Ediacaran fauna, soft-bodied bottom-dwelling fish, and at shellfish that are among the earliest shelled organisms preserved in the rocks. The fossils are all found in rock strata dated as about 600–550 million years old, near the beginning of the Palaeozoic era of geological time. It seems reasonable to suppose that similar fauna preserved in rocks of the same age as one another were living in close proximity at that time, and comparison of the fossils from this time show two distinct patterns.

He reports that when the continents are rearranged so as to bring the "Ediacaran" rocks together and the "oldest shelly fauna" rocks together, two geographically distinct "provinces" emerge. The home of the benthic (bottom-dwelling) Ediacarans

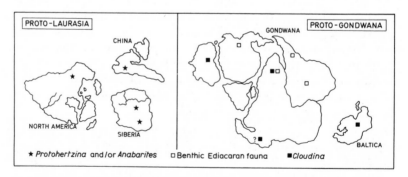

Figure 21.5 *Rocks of the same age from different parts of the world contain different fossil remains. When continents are arranged so that early Palaeozoic strata containing similar remains are next to each other a picture of two supercontinents which existed some 550 million years ago, emerges*

corresponds to certain areas of Australia, Africa and the Baltic. A problematic fossil, which McMenamin also considers a member of the Ediacaran benthic community, is found in Antarctica and South America, as well as in Africa where it occurs with typical benthic Ediacarans.

The very first creatures with shells are restricted to other areas, primarily in China, North America and Siberia. None of these continents contains any record of benthic Ediacarans.

When the continents are arranged on a map to bring the regions with similar remains together, the result is a map dominated by two supercontinents, Laurasia and Gondwana, just as the palaeomagnetic evidence suggests (*Geology*, vol. 10, p. 293).

This evidence of "faunal provincialisation" lends weight to the idea that Pangaea fragmented well before the beginning of the Palaeozoic, with rifting and continental movements under way about 800 million years ago. It confirms the picture of the two supercontinents pieced together from other data – and it provides new insights into the period of explosive diversification of animal life in the early Palaeozoic, a period of great change which may well have been stimulated by the dramatic changes in the geographical environment.

"MONITOR", 23 September, 1982

FOSSIL INSECTS SHOW THE WORLD AS IT WAS

Fossils of arthropods, and possibly insects, found near Gilboa in New York State, promise to shed light on an important biological event in the far-distant past – the evolution of the earliest land animals and plants in late Silurian and early to middle Devonian times, around 420 to 375 million years ago. At about the same time, fish were moving from the marine environment into fresh waters and extensive land masses were forming as continents merged.

The world looked very different from the way it does today. As was clear at a Royal Society meeting held early this month to discuss this crucial period, opinion still differs about where some of the continents were at the time. In particular, how near was North America and parts, or all, of Eurasia to the great southern supercontinent of Gondwana, which comprised South America, Africa, Arabia, Antarctica, Australasia and peninsular India? It

seems fairly certain that Great Britain – and the former delta
region where the Gilboa site is now – were somewhere near the
equator; the South Pole lay in or near Southern Africa.

The land animals of around 400 million years ago were
arthropods, such as arachnids (spiders, mites and such-like)
myriapods (centipedes and millipedes), and various extinct
groups. Ancestors of these first land-dwelling (terrestrial) arthropods would have been aquatic but their identity is uncertain; there
may have been several separate invasions of the land by different
arthropods. Although this period around 400 million years ago
was a time of great biological change there were no marked
"bursts" of extinctions of life.

The earliest unequivocal terrestrial arthropods (millipede-like
myriapods) are known from late Silurian rocks in Scotland but
evidence for them is scanty. Early Devonian terrestrial arthropods
have been found at Rhynie in Scotland and Alken an der Mosel in
West Germany. At about 376–379 million years old, the Gilboa
arthropods are slightly younger than these European fossils but
they are much better preserved and they represent a greater variety
of forms, as William Shear of Hampden-Sydney College and
colleagues show in a recent paper in *Science* (vol. 224, p. 492).

Among the Gilboa arthropods are a spider, a well-preserved
mite, various other early arachnids (trigonotarbids) and centipedes. Modern representatives of these forms live in crevices,

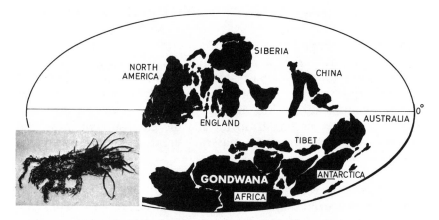

Figure 21.6   *A primitive mite (inset), one of the early land
animals that lived around 377 million years ago, probably
near the equator*

under stones and among plant litter so there is good reason to believe that the Gilboa fauna was also cryptozoic; hiding in this way would have prevented loss of water.

What could be an early insect belonging to the primitive order Archaeognatha (*archaeo* = ancient; *gnatha* = jaw), or jumping bristletails, was also found in the Gilboa rocks. It is represented by part of a head with a compound eye showing hexagonal facets. Similar forms are still found today; they are sometimes grouped with the bristletails proper in the class Thysanura. If its identification is confirmed the fossil bristletail could be the oldest insect known. Previously the earliest records of insects – if collembolans or springtails are excluded – were from rocks of Carboniferous age some 50 million years younger than those at Gilboa. There are collembolans about 390 million years old from the later part of the early Devonian, but these wingless arthropods are now often placed in a different class from other insects.

One of the features of the Gilboa site is the preservation of hundreds of pieces of arthropod cuticle amongst the more easily identifiable body parts and whole specimens. The palaeontologists discovered the cuticle fragments when they were treating rock samples with hydrofluoric acid to obtain fossils of early land plants called lycopods (club mosses) which are abundant at Gilboa. According to Shear and his colleagues these pieces of cuticle are "flexible and tough enough to be removed from the acid bath with forceps, washed, and mounted on slides". They show sense organs, fine hairlike structures (setae) and minute sculpturing of the surface. It is very likely that such fragments have been missed by palaeontologists in the past.

"MONITOR", *31 May, 1984*

22

Getting down to ocean-floor geology

PETER STUBBS

Geophysicists, at a Royal Astronomical Society meeting last week, reported on their latest results from mapping the slow oozing of the ocean floors.

The discovery that the ocean floors are ponderously spreading outwards from the ridges that form their spines has led to an entirely new and highly exciting kind of geology. A variety of patterns of oozing are coming to light and the deep sea bottom reveals a complex of structures arising from combinations of ribbon-like flow, twisting and sheering. Unlike the rather more conservative Earth movements that take place within continents, all these plastic processes result from the production of new viscous material inside the oceans. In the past 18 months or so since it became clearly apparent that the ocean floors act like huge conveyor belts, transporting the continents across the globe (see "The mechanism of continental drift", by Dr Peter Stubbs, *New Scientist*, vol. 32, p. 616), oceanographers and geophysicists have vigorously tilled the seas by magnetic mapping, the key technique for disinterring what the oceans have been up to. Last Friday the Royal Astronomical Society held a meeting to discuss the latest fruits of this productive husbandry.

The particular clue which led researchers to this new marine exploration was the finding that the magnetic pattern of the ocean floor is typically striated, the magnetic anomalies forming long, narrow strips. These, it has been suggested, may mark the alternate periods during which the Earth's magnetic field was either normally or reversely directed. Whether or not this is the true explanation, there is ample evidence to show that the magnetic lineations are all lined up essentially parallel to the mid-oceanic ridges. Moreover, they form exceptionally symmetrical patterns, in many cases, with a set of striations on one side of a ridge mirrored almost exactly on the opposite side. It seems hard to find any explanation to fit the facts bar that of a spreading ocean floor, in which lavas extruded along the ridge freeze into themselves the prevailing magnetic field directions.

The magnetic mapping to date has been performed either by towing a magnetometer behind a ship, or by flown instruments. To some extent the underlying topography of the sea floor must affect the records and it is clearly desirable to find out how big this effect is. Moreover, since the water over those areas is usually at least two km deep, present methods make the resolution of the detail impossible. Are the minor magnetic peaks and troughs of the magnetometer trace the result of lava flows, intrusive sheets of molten rock-like dykes, or simply due to variations in the extent to which these structures have become magnetised?

At Scripps Oceanographic Institute, La Jolla, California, geophysicists have perfected a means of answering some of these questions. Unravelling deep-sea geology can, in part, be done by drilling as in the Joint Oceanographic Institutions' Deep Earth Sampling (JOIDES) programme now about to begin (see *New Scientist,* vol. 38, p. 386). But valuable data can also be obtained more cheaply by towing equipment along near the sea bottom. FISH, devised at Scripps, is the first comprehensive gadget of this type. Dr R. G. Mason of Imperial College, London, one of the initial workers to realise the potentialities of magnetic mapping in oceanography, told the RAS gathering how, in its first two surveys, FISH has advanced ocean-floor geology by one more substantial stride.

FISH is a submarine bundle of echo-sounders clustered about a sensitive magnetometer. It has a wide-beam echo-sounder pointing upwards to measure its depth by signals bounced off the sea surface; a narrow-beam, downwards echo-sounder to map the bottom topography; a seismic profiler to bounce sounds off the hard rock underlying the bottom sediments; a sideways sonar to pick out mounds, cliffs and the like; a navigational bleeper to locate its position accurately relative to surface transponders; and the magnetometer.

With this powerful tool its designers can map the ocean floors to an entirely new order of accuracy. The FISH is towed along at one or two knots, kept as closely as possible to 200 m above the bottom. At a rate of two readings a second it provides a set of measurements about every 15 cm.

The two surveys made to date encompass part of the Gorda Ridge, a well-mapped area of the north-east Pacific characterised by distinctive magnetic lineations; and part of the floor near the Murray Ridge south-east of San Francisco. The first of these was conducted slap across the central trough of the Gorda Ridge, a rift that is the common feature of nearly all ocean ridges. Within the 30 km wide negative magnetic anomaly, which is all that surface surveys reveal, there is a wealth of finer detail. First, the survey shows that while the topography correlates in a positive fashion with the magnetics at the trough centre, farther out there is a clear-cut disparity between the two. Here, it seems, is direct evidence that the rocks really are reversely magnetised as postulated. Small fluctuations in magnetic intensity at the centre suggest that one mechanism by which the ocean floor spreading takes place may be by stretching and the intrusion of multiple dykes – a process that appears to be going on at the moment in Iceland, which lies across the Mid-Atlantic Ridge.

The topography, shown up in detail by FISH, reveals the central rift of the Gorda Ridge as a succession of scarp and dip slopes arranged as rift valleys within rift valleys.

The second FISH survey has disclosed a similar pattern across the Murray Ridge where four or five small anomalies appear in the central trough which persist along the axis of the ridge for 20 or 30 km. Off-axis traverse allowed a 500 sq. km area to be contoured at intervals of 20 metres and plainly showed that the topography was arranged, like the magnetism, in long thin strips. Mapping on this scale is tedious work and it is unlikely that the technique will ever be employed for world-wide surveys. But for unravelling the nature of the features typical of the ocean floors there is no doubt now about the value of towable devices such as FISH. If enough detail is forthcoming about intrusive and extrusive rocks along these oceanic troughs, it may prove possible to make comparisons with, say, Iceland where the stretching rate can be gauged fairly accurately, and hence deduce an independent value for the rate at which the ocean floors are spreading – currently reckoned to be a few cm a year.

Other contributors to last Friday's meeting described the results of more conventional magnetometry – but they were nonetheless startling for that. One or two geologists in the past have suggested that the Pyrenees represent a continental crumpling movement resulting from a rotation of Spain anticlockwise, with the consequent tearing open of the Bay of Biscay which is a feature of oceanic depth. Rock magnetism studies on land bear out the idea, indicating that the rotation has amounted to 32° since Triassic times 200 million years ago. The final 22° of this movement seem to have happened in the past 50 million years. Now Dr D. H. Matthews of Cambridge University has mapped the Bay itself and finds a convincing fanshaped array of magnetic anomalies spread over 35°. This is totally at variance with the predominantly south-west by north-east trend along the English Channel. There does not seem to be any central ridge in this case and Dr Matthews thinks his results may be another reflection of igneous rocks intruded into tension cracks. But what made Spain rotate?

Widespread faulting along the Carlsberg Ridge in the Indian Ocean has a pattern, according to Dr A. S. Laughton of the National Institute of Oceanography, which suggests an even bigger rotation – that of the whole of Africa about a point near the centre of its north coast and amounting to some four degrees per 10 million years.

Extensive mapping of the main ocean ridges is continuing with enthusiasm according to Dr J. R. Heirtzler of Columbia University's Lamont Geological Observatory. American workers hope eventually to produce a complete "isochron" (lines of equal age) map of the world's oceans. Areas, where the pattern of oozing is complex, include Iceland and the related Norwegian Basin, around the Galapagos Islands, and a huge sheared flexure south of the Aleutian Islands.

30 May, 1968

23

Dating the spreading sea floor

DAVID FISHER

The recent sea floor spreading hypothesis implies that the rocks farthest from the crests of oceanic ridges should also be the oldest. Radiometric ages of deep-sea rocks now appear unreliable; but fission-track dating may supply the final proof of this attractive theory. Meanwhile, isotope analysis can find a new application in measuring rare-gas abundances deep within the Earth.

The elegant new hypothesis of sea-floor spreading has provided fresh impetus to our search for methods to measure the age of the ocean bottom. Precise dates of the times of eruption of ocean-floor rocks now located at particular distances from the active mid-ocean "rises" may provide the final proof or disproof of the sea-floor spreading concept. In this concept the active rises are visualised as the sites of upwelling deep Earth material. The material erupts along the crest of the rises and from there pushes out to form a "spreading sea floor", as illustrated in Figure 23.1. The age of the sea floor will thus be greatest farthest from the rises, and youngest close to them. A well-mapped-out programme of age determinations will measure directly both the rates of spreading in various parts of the world's oceans and the directions of such spreading.

Radiometric age determinations on sub-aerial rocks, while relatively new in the history of science, have already become well established and, in many instances, routine. Of these, the potassium–argon method, in which use is made of the decay of potassium-40 into argon-40, has proven to be by far the most reliable and is applicable to the age region – 1 to 100 million years – of interest in this problem.

The first thing to do in attempting to date deep-sea rocks was therefore to apply this "classical" method of analysis.

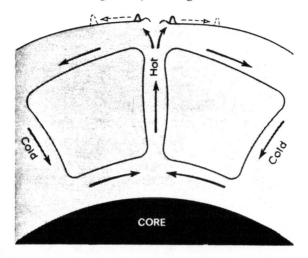

Figure 23.1 *Deep Earth material appears to upwell along the crest of an active mid-ocean rise, and the sea floor then spreads away from the rise*

The results proved surprising. The Cobb Seamount in the north-east Pacific showed ages of 19, 35, 0.5, and 1.6 million years, as measured by two different groups. Replicate analyses, made by a third team on another seamount, gave ages ranging from 2 to 200 million years. Something was certainly wrong, but with no clue to the true ages of the rock samples it was not clear in which direction to search further.

At this stage we obtained samples from the mid-Pacific that were ideally suited not only to a radiometric age investigation pertinent to the concept of sea floor spreading, but also to a study of what exactly was wrong with ocean-floor dating *per se* (Figure 23.2). A broad outcrop of rock tens of kilometres wide and hundreds of kilometres long was found to lie along the crest of the mid-Pacific ridge (Figure 23.3). According to the sea-floor spreading concept this ridge is thought to be a topographic extension of an upwelling limb of a convection cell in the Earth's mantle, and therefore this rock should be recent – certainly less than 1 million years old. There is no sediment cover to the rock plateau, again indicating a young age. On the other hand, if the seamounts, numbered 20, 23, 25, and 50 on Figure 23.2, now located at distances ranging from 500 to 1500 km from the crest of the rise, actually originated on the crest and drifted to their present

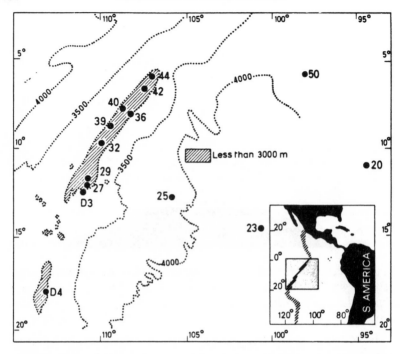

Figure 23.2 *Bathymetric map of a portion of the East Pacific Rise, showing locations of samples dredged for dating purposes*

positions along with the spreading sea floor, they should show much greater ages. Drifting at the currently estimated rates of about four cm per year, the ages of these seamounts should increase to a maximum of about 40 million years for the farthest one.

The potassium–argon ages that John Funkhouser, of the State University of New York, Stony Brook, and I measured for these rocks, however, did not fit those expectations. About half the rock samples dredged from along the crests of the rise did indeed yield ages of less than 1 million years, but others showed ages ranging up to nearly 700 million years. At first glance these high ages seemed to be exciting evidence against the sea-floor spreading hypothesis, but upon a little reflection they turned out to be totally unacceptable as real ages, and must instead be interpreted in a different light.

First, visual observations of the plateau of rock from which all the samples were dredged revealed it to be a continuous formation. It does not seem reasonable that some of the rocks in this formation should be less than 1 million years old and others over 600 million years old. Both the high and low ages cannot be correct. Second, as I have mentioned, the lack of sediment along the entire plateau indicates that it is very young. Rocks many millions of years in age would be covered by tens of metres of sediment, as they are in other parts of the world's oceans. It therefore seems likely that the excessive potassium–argon ages are indicative of large amounts of "excess" argon-40.

In order to prove this contention, it was necessary either to date rocks of known age, or to apply another method of dating to those rocks showing "excess" argon. Both these experiments have now been carried out and have yielded positive results.

G. Dalrymple and H. Moore, and independently C. S. Noble and J. J. Naughton, carried out a series of potassium–argon age determinations on submarine lavas known to be much less than 1 million years. Several of the samples displayed ages ranging up to about 40 million years, supplying definite proof of the occurrence of "excess" argon in submarine rocks.

Figure 23.3 *Part of a basalt pavement on the ocean floor along the crest of the East Pacific Rise (Credit: School of Marine and Atmospheric Science, Division of Marine Geology and Geophysics, University of Miami)*

At the same time I was dating by the fission-track method those rocks from the East Pacific Rise which we believed showed evidence of "excess" argon. Briefly, this method makes use of the fact that uranium-238 (which is present in the submarine rocks in trace amounts) undergoes spontaneous fission on a very long time-scale. In the fission process, two massive and heavily charged particles are shot away from each other with tremendous energy, depositing about 200 MeV along a channel approximately 10 to 20 micrometres in length, which becomes microscopically visible upon chemical etching. The density of these spontaneous tracks is proportional to both the age and the uranium content of the rock. Tracks induced by a subsequent neutron irradiation are proportional only to the uranium content; therefore it is possible to calculate the age of the sample from the ratio of induced tracks to spontaneous tracks. All this has been worked out in a beautiful series of papers by the General Electric Schenectady group: P. B. Price (now at Berkeley), R. L. Fleischer, and R. M. Walker (now at Washington University).

Application of the technique to the East Pacific Rise samples showing "excess" argon has completely reinforced our interpretation. All the samples along the crest of the ridge that were dated have fission-track ages of less than 1 million years. The fission-track ages can be accepted with confidence as a good approximation to the true age of eruption of these basalts. There is no way for fission tracks formed previous to eruption to be maintained in the volcanic glasses, as previous studies show that the tracks anneal out and disappear rapidly at elevated temperatures. It does not seem likely that uranium can be leached out of the glasses without disrupting the structure and therefore the fossil tracks. Both these effects would give excessively old ages; the only mechanism so far discovered for the fission-track method to give anomalously young ages is that in which fossil fission tracks are annealed out at elevated temperatures. This is sometimes a severe problem in sub-aerial minerals, but not in submarine dating where the ambient water temperatures are near freezing.

Relatively large amounts of "excess" argon-40 therefore seem to be a common component of deep-sea volcanic rocks. The general features of a model to explain this occurrence do not seem too difficult to visualise. In the sub-aerial case, the model by which one assigns potassium–argon ages to igneous rocks is roughly the following: At some stage the rock material, containing potassium which is constantly decaying to argon-40, is molten at depth within the Earth. This magma erupts and, as the internal pressure is released, the argon-40 produced up to that time diffuses out. The rock then cools to a temperature at which it begins to retain argon quantitatively, and it is this "event" which we date. In the deep-sea case a similar model must hold true. Here, however, the rock is erupting, not into the air, but into water at practically 0° C and many kilobars of pressure. In this situation we can imagine two different cases. If the ascent of the molten material from within the Earth to the surface is slow, then as it rises it will begin to cool and crystals will begin to form. As they crystallise they push argon, which is a large atom, out of their structure. If, by the time the rock material reaches the surface of the ocean floor, it is mainly crystalline, the crystals will be free of argon which formed prior to eruption and will be suitable for dating. If the magma rises rapidly, however, crystals will not have time to form and as the magma bursts out onto the ocean floor it encounters the nearly 0° C sea water at high pressures. The rock is chilled quickly to a glass and the pre-eruption argon does not have time to escape before it is trapped as "excess" argon in this glassy matrix.

These observations, while diminishing the value of potassium/argon dating for deep-sea rocks, open new areas of research in rare-gas abundances of the deep Earth. To illustrate this, we measured a helium-4 abundance of about 1300×10^{-8} cu. cm/g in the glass crust of sample 44, one of the samples with large amounts of "excess" argon. Our fission-track studies showed a uranium abundance of about 0.1 ppm in this glass. Using the upper limit of 1 million years for the age of this sample, set by the fission-track work, we calculated that over 99 per cent of the helium-4 is also "excess", i.e., of radiogenic origin but produced prior to the eruption and subsequent cooling of the glass.

This study gives, for sample 44, a helium-4/argon-40 excess ratio of about 13. Such a ratio is approximately the maximum to be expected from estimated ratios in the Earth's mantle of uranium/potassium of approximately 10^{-4}, and thorium/uranium about 3.6. Since the radiogenic helium-4/argon-40 ratio decreases with time, this indicates two things: first, that the parent material of these rocks has operated as a closed system (with respect to the rare gases) for only a short time compared to the age of the Earth; and, second, that during this time there has been insignificant degassing of the rare gases, so that the measured values indicate the rare-gas atmosphere ambient in the parent material.

These observations open up the possibility of studying rare gas abundances and composition deep within the mantle. Comparison of the isotopic and elemental abundances of such rare gases (particularly of the heavy gases krypton and xenon) with corresponding nuclide abundances in meteorites bear on the early history of the Solar System. Studies of this type will be most useful if the rare-gas abundances turn out to be rather uniform from location to location, indicating a homogeneous mantle concentration. On the other hand, if large variations are observed it may be possible to study local mantle concentrations of uranium, thorium, and potassium by analysing the helium and argon decay products. Hopefully, this would give us the true mantle abundances of these important elements rather than merely their abundances in mantle-derived rocks, as is the case at the moment.

Even at present, however, the potassium–argon ages obtained on deep-sea rocks do have some value as they may be considered to be valid upper limits to the age of eruption, since we have no reason to suspect diffusion loss of argon. In particular, the very cold ambient water temperatures that inhibit loss of the pre-eruption argon will also inhibit diffusive loss of post-eruption argon. Then the ages we obtained for seamounts at varying distances from the East Pacific Rise are of validity in relation to the concept of the sea-floor spreading. As I pointed out earlier these ages should vary with distance from the rise, reaching a maximum of something like 40 million years for the seamounts under consideration. In contrast, the data show all the seamounts to be approximately the same age – about 3–4 million years old. There is no increase in age with distance, and the total age measured is nowhere near what we would have expected.

These data should not be used to argue against the general concept of sea floor spreading, but they do indicate at least one particular correction. Since the three seamounts we studied are all too young to have originated on the crest of the rise and drifted to their present positions, we have to give up the idea that the open sea floors are rather devoid of volcanism and that most submarine volcanism takes place along the crest of the active ridges, as originally envisaged. Instead it would seem, from these limited data, that submarine volcanism is a rather general feature of the sea floor.

In order to test the more general features of the sea floor spreading model, we must date the lava flows which form the actual sea floor bottom and not the sea mountains which spring up from this bottom. During the spreading these flows are buried beneath sediments and cannot be sampled by normal techniques. The JOIDES Deep-Earth Sampling Program, however, is currently involved in obtaining deep cores from the sea floor. Several of these have already dug through to the underlying basalt flows, and we can look forward to the availability of many datable and pertinent samples in the near future. The problem of dating them correctly remains, and fission-track analysis seems to be the most likely method. A few results have been obtained only recently (from Atlantic Ocean rocks by the General Electric group and F. Aumento of Canada; and on a Pacific abyssal hill by B. P. Luyendyk and myself) which indicate agreement with the spreading hypothesis, and which in fact lead to a spreading rate in agreement with theory – but we must appreciate the small number of samples involved. Future fission-track studies, with a statistically significant number of samples, will delineate the spreading rates and directions more exactly.

22 October, 1969

24

Hotwater life thrives around submarine volcanoes

The Galapagos Rift Zone is an east-west ocean-floor rift that branches off the East Pacific Rise, the main sea-floor spreading axis of the Pacific. It runs just north of the equator and the Galapagos Islands famous for Darwin's observations. In February and March scientists from Scripps Institution of Oceanography, San Diego, California, spent some time exploring the underwater nature of the rift zone. They had previously discovered along this old volcanic region periodic hydrothermal springs. Closer examination employing the celebrated submersible craft *Alvin* now reveals the presence of abundant faunal assemblages specifically associated with these plumes of hot water.

Submarine photography had shown communities of white mussels living on mineral-encrusted mounds and around sea-floor volcanic vents. But the *Alvin* trip, to depths of 2500 metres, disclosed also big clams, 10-inch sea anemones, and large hoards of yellow crabs.

Where the hydrothermal sources were absent, and water temperatures at the norm of around 2° C, marine animals were scarce. But the Scripps oceanographers, headed by Dr John Corliss, discovered three hot springs with what they term faunal "clambakes" thriving in the warm water at some 11° C. A fourth community appeared to have died at the site of a now inactive spring. According to geologist Kathleen Crane, the first woman scientist to dive to such depths in the *Alvin*, the clambakes offered "... strange scenes, some with invertebrates that resembled dandelions and others with cobweb-like objects draped over volcanic rocks. The animals moved slowly in the spa-like water that created a shimmering effect as it sprang from vents in the slime-covered lava."

The *Alvin* towed an instrument package to photograph the sea

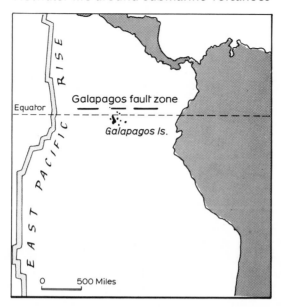

Figure 24.1

floor, measure temperatures, pressures, and light transmission, and to sample the water, particularly that issuing from the hot springs. Later, on board the *R V Melville* and ashore, researchers from the Scripps Isotope Laboratory analysed the water samples for primordial gases, mainly helium isotopes, in a search for the origin of the hot water. Is it so-called "meteoric" water emerging at the Earth's surface for the first time ever?

Another major problem concerns the source of nutriment for these strange abyssal creatures. Francisco Vidal is now studying bacterial cultures from the water samples to look for anaerobic hydrogen-sulphide metabolising organisms which may provide the primary production at the start of the submarine food chain. Such bacteria may be living in the slime that coats rocks around the hydrothermal sources.

An exciting prospect voiced by Kathleen Crane is the chance of finding new species among the clambake faunas.

"MONITOR", 21 April, 1977

25

Challenge of the sea bed

JOE CANN

For more than 10 years the drilling ship Glomar Challenger *has been perforating the floors of the world's oceans as part of an elaborate project to explore one of Earth's last frontiers. What have the geologists found so far, and why are they keen to continue such a costly venture?*

A recent party in the small Mexican port of Mazatlan celebrated the tenth anniversary of the Deep Sea Drilling Project (DSDP). Dr Archie Douglas, an organic geochemist from Newcastle University and UK participant in leg 63 of the project, was there. He says that after eight weeks at sea on a dry ship any party would have been most welcome; and though the Margueritas may not have been mixed to the highest standards, this was more than compensated for by their being there at all, along with dancing to midnight, and a buffet supper with salad and fresh fruit.

Why the celebration? For those new on the deep sea drilling scene, perhaps I should recapitulate that the project funds the ship *Glomar Challenger* to drill holes in the sea bed for scientific purposes. The ship – a 10 000 tonne motorised soap-dish with a hole in the middle – positions itself over a sonar beacon on the ocean floor, and runs the drill string down into the sediments, volcanic rocks or whatever that lie beneath it (*New Scientist*, vol. 79, p. 10). Since 1975 the project has been an international one, with France, West Germany, Japan, the USSR and the UK paying an annual subscription of about $1 million each. The member states also provide representatives on the advisory committees, participants on the ship, and research on the rocks brought back from the ocean crust.

Among its early successes the project demonstrated that the ages

Figure 25.1 *The derrick (above) and sections of drill pipe on board* Challenger *(overleaf) (Credit: C. Sutton (above)/ Gary Pritchard (overleaf))*

Figure 25.1 *(cont.)*

of sediments overlying basalt basement in the South Atlantic corresponded closely with the ages deduced from the magnetic anomalies (regions which show departures from the predicted value of the Earth's magnetic field). This was compelling evidence for continental drift. The drilling also revolutionised our ideas of the geology of the ocean floor, especially in the interpretation of the sediments where, to begin with, unexpected quantities of *chert* (a sediment composed of hard silica, SiO_2; flint is a variety) were found. More recently, the sedimentary record of the ocean floor has provided the principal impetus to the study of palaeo-oceanography (looking at past patterns of ocean circulation) and of global climate. In this article I shall describe some of the recent success of the project, some problems it has raised, and the way plans are shaping for the future.

The *hot spot* is one idea that has had great success in the past few years. It contends that any chain of volcanoes on the Earth's surface is produced by some anomaly inside the Earth which stays fixed relative to the Earth's deep interior. In some pictures this anomaly takes the form of a narrow jet, or plume, of rising mantle originating perhaps as deep as the core. Chains of volcanoes therefore provide direct evidence of the motion of plates relative to the Earth's interior, and of the history of plate movement through time.

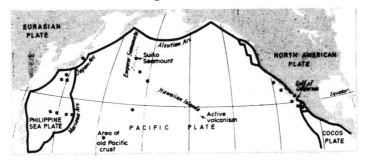

Figure 25.2 *Plate boundaries in the region of the North Pacific. Stars indicate areas recently drilled by* Glomar Challenger

The Hawaiian chain of islands provides a classic example of a string of hot spot volcanoes. Recent activity is concentrated at the eastern end of the chain, the volcanoes being successively older towards the west. Beneath the sea a string of seamounts (elevated, but submerged regions of the sea-floor) continues the line westwards as far as longitude 165° E, where it meets the Emperor seamounts, which form a chain running north to the Aleutian island arc (Figure 25.2). Does the change in the direction of the chain mark a change in the direction of motion of the Pacific plate? Are the Emperor Seamounts and the Hawaiian chain products of the same hot spot? Leg 55 set out to test these questions by drilling a deep hole in one of the Emperor Seamounts. The sequence of lava types in the hole (in Suiko Seamount) was the same as the characteristic sequence found again and again in the Hawaiian islands. These lavas had clearly been erupted on land, submerging later as the island sank below sea-level. In addition, the magnetisation of the lavas showed that they had originally been erupted at a latitude close to that of Hawaii itself, 19° N, though they now lie at a latitude of 43° N. So this drilling, masterminded by Dale Jackson of the US Geological Survey, who died a few months ago, gave a well-directed boost to hot spot ideas.

Under Iceland there ought to be another hot spot, this time more or less beneath the axis of the Mid-Atlantic Ridge, where it cuts through Iceland. To the south of Iceland, Jean-Guy Schilling, of the University of Rhode Island, found a gradient in the composition of the lavas that he dredged there. He attributed this to mixing of lavas derived from the hot spot activity, which are rich in radioactive and other melt-loving elements, with others produced during normal ocean-floor spreading. The hot-spot lavas would become progressively diluted with ocean floor lavas further from Iceland. Leg 49 drilled three holes in progressively older crust, half-way down the gradient, and found that nowhere was the picture so simple. Each site revealed virtually the whole range of composition and there seemed to be no clear-cut progression with age. Instead, intensive geochemical work on the leg 49 rocks, spearheaded by a splendidly amicable collaboration between laboratories in France and the UK, suggested a picture which involved inhomogeneities in the composition of the mantle on a whole variety of scales. This situation could be the result of migrating fluids in the mantle being stirred by a "treacly" convection, and could hold some important information about the evolution of the Earth through geological time.

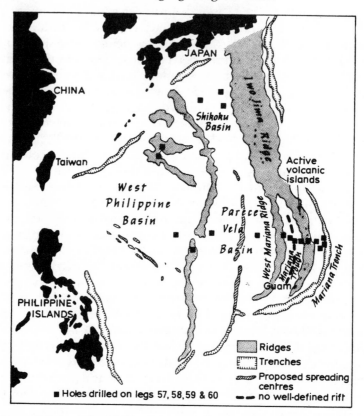

Figure 25.3 *The Philippine sea, showing trenches, arcs and inter-arc basins*

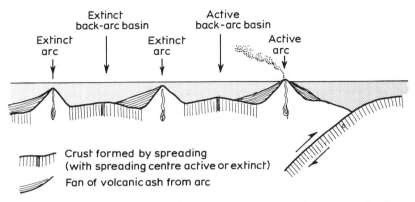

Figure 25.4 *The model of back-arc spreading in which successively older island arcs are separated by successively older spread basins*

More recently in five successive legs (56–60), *Glomar Challenger* drilled in the volcanic arc complexes of the western Pacific. The chains of island arcs there form festoons sometimes two or three deep between the continent of Asia and the deep Pacific floor. Each arc consists of a chain of island volcanoes outside which lie deep trenches. These trenches are the site of subduction of the Pacific plate, where it returns to the mantle, and its descent is marked by a plane of earthquakes which dips beneath the volcanoes of the island arc. Depth contour maps show that behind each arc is a series of ridges and troughs parallel to each other and the arc (Figure 25.3).

Put simply, the drilling in this region resulted in the triumphant confirmation of one hypothesis and the destruction of another, at least in its simplest form. The winning hypothesis was that of *back-arc spreading*, first put forward in 1970 by Dan Karig, a bearded, ascetic geophysicist from Scripps Institution of Oceanography. Karig suggested that the arc periodically split in two, with the half nearer the continent becoming volcanically extinct, and gradually spreading away from the still active outer part, so creating new ocean crust in a *back-arc basin* between them (Figure 25.4). Thus each consecutive ridge should be a progressively older half-arc, and each trough a progressively older spread basin. The drilling confirmed this picture piece by piece, and at the same time dated the successive splitting episodes. When a hole was drilled on a ridge, volcanic sediments typical of the island arcs were recovered, at depths of up to 1 km. In the western parts of troughs, away from the arc deep-sea sediments were found to overlie basalt basement formed by spreading; in the eastern parts, the drill pierced thick fans of volcanic sediment from the active arcs.

The hypothesis that lost was that of the *accretionary prism*. According to this, the zone between the volcanic arc and the trench ought to be composed of slices of oceanic crust and oceanic sediment detached from the down-going slab and stacked one against the other to form a pile dipping under the arc (Figure 25.5). Structures very like this pile appeared on some seismic reflection profiles crossing the arc-trench zones. It all made a very good story, but, the drilling did not confirm it.

What the drilling found, instead of the *pelagic* sediments predicted (precipitated from the open ocean), was a lot of sediment clearly derived from the volcanic arc. Instead of showing a progressive shallowing with time as the slices piled up, one site off Japan showed a progressive deepening with time. The hole bottomed in the beach deposit of a fossil landmass, though it was drilled in water nearly 1000 m deep. In some holes, a particular and characteristic type of magnesium rich lava called *boninite* occurred. This kind of lava has never been found on the ocean floor, but has now turned up in four or five places in island arcs. It is another indicator that whatever the origin of the rocks in this

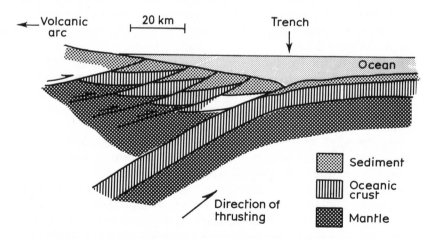

Figure 25.5 *The accretionary prism model in which slices of ocean crust and sediments are stacked up between arcs and trenches*

zone, they are indigenous to the arcs and not scraped-off pieces of ocean crust.

Much of the drilling on the following legs (61–63) was aimed at understanding the sediments of the Pacific Ocean, particularly in relation to changes in the circulation pattern of the surface waters and the northward migration of the Pacific plate on which the sediments are laid down. By the nature of things, palaeontologists (who study ancient organisms) are not hasty men; they consider and cautiously change their minds back and forth. So, settled conclusions from these legs are not likely to arise for some while, but it is possible to give an outline of the general results.

The method by which the problem is tackled is essentially to use fossil organisms, especially micro-fossils, to give indications not only of age, but also of surface water temperature and of past water circulation systems. For this to work well, as many components as possible of the original faunal and floral assemblage should be preserved. This causes problems in the deep ocean basins; the crust there is old enough for it to have sunk to depths at which the cold bottom water can dissolve the calcareous skeletons of surface-living organisms. In such deep water very little of the surface skeletal debris survives to be buried – in some parts of the world the sediment is entirely barren of fossils. So to get the whole story it is important to find areas where the water was once shallow, such as mid-oceanic ridges, or the enigmatic swells or rises that sometimes make the ocean floor unusually shallow (perhaps the result of hot spot activity of the kind I talked about above). The picture is made even more complicated by the equally enigmatic gaps in the sedimentary record. These extend, as far as we can tell, over quite large areas at particular times. Do they represent periods of intense current activity that prevented sediments from settling and eroded away some that had? Or have they a more complex origin?

Leg 61 was occupied with drilling a single deep hole in an area (the Nauru Basin) where magnetic evidence suggested that the crust was very old, perhaps mid-Jurassic in age (about 170 million years old), certainly older than any other Pacific crust drilled so far. We wanted to know what the Pacific – presumably even larger than it is now – used to be like in one of the few scraps of crust that have escaped being subducted, and palaeontologists and sedimentologists fought to get on the ship. In the end no Jurassic sediment was found at all. The drill went down through sediments getting progressively older as far as the mid-Cretaceous (about

100 million years old), when it suddenly hit basalt lava. The *Challenger* team drilled through 500 m of this, occasionally picking up tantalising fragments of sediment gradually increasing in age to Lower Cretaceous (about 130 million years old), but, in the end, time ran out and the hole had to be abandoned for other things. Where did all the basalt come from? Was the Jurassic sediment there at the bottom somewhere, perhaps just below where the drilling stopped? Certainly this site raised many more problems than it solved. As for legs 62 and 63, interpretation of their results awaits further study onshore.

The plans for the next two years beyond the present legs are gradually firming up, although American funding for drilling beginning in October will not be decided until August. If this is agreed, a varied programme is proposed for the Atlantic area in general, including studies of the continental margins of Europe and North America. One of the highlights is likely to be the use of the new piston corer for examining the record of climatic changes contained in sediments. Another important aim is to improve understanding of potentially petroliferous continental margins.

Looking forward beyond 1981, the future becomes very clouded. One possibility is to use the CIA-commissioned deep-ocean recovery ship *Glomar Explorer* for continuing the drilling programme (*New Scientist*, vol. 79, p. 10). At present we are in the throes of design studies, feasibility studies and all the rest of the (remarkably expensive) preliminaries to writing a proposal naming objectives and realistic sums of money. In order to start this, the planning committee at its last meeting strung together a proposed programme of drilling of the kind at which *Glomar Explorer* is expected to excel. The programme to some extent wrote itself, due to the large number of constraints: for example, *Glomar Explorer* is too large for the Panama Canal, so that if it is to get through to the Atlantic it has to pass through the Southern Ocean, preferably during the southern summer. So far the programme for *Explorer* envisages an initial phase of Pacific drilling, picking up some sites out of reach of *Challenger* at present; a passage through the Southern Ocean, drilling sites at latitudes too far south for *Challenger;* and a long spell drilling deep holes in continental margins in the Atlantic. This phase of margin drilling would occupy four or five years, and would be followed by a return to the Pacific through the Indian Ocean.

However, this is merely an outline to start the planning process off, and it could be substantially modified when the costs become clearer. For example, the United States Geological Survey may want to become more closely involved with the whole thing, and attract much more of the drilling into US waters. Or, the foreign participants might not feel that they want to stay in the project, or indeed might not be welcomed. Certainly the drilling is going to be a very different operation than in the past. Not only will *Explorer* be more expensive to operate, but some of the very deep continental margin holes may take as long as a year to drill. So the cost of the piece of core could be truly immense.

You might well ask whether any piece of rock, whatever its splendid nature, could possibly be worth such an investment? I think there are two answers to this question. One is that if a piece of rock led us to a new geological environment for recovering fossil fuels, then it would well be worth it. The other is that similar questions were raised originally about the value of any sort of scientific drilling in the ocean basins. I am sure that the kind of results I outlined above shows that these objections were not well founded, and the same might well turn out to be true for drilling with *Glomar Explorer*. Certainly the results of drilling so far have changed our whole view of the geology of both ocean basins and the rest of the Earth, and will continue to do so as the data are assimilated.

7 June, 1979

26

Plate tectonics: where is it going?

TONY WATTS

The concept of plate tectonics has revolutionised the Earth sciences; what happens next in the chain of events that began with the birth of Alfred Wegener, 100 years ago this week?

The concept of plate tectonics views the outer layer of the Earth — the lithosphere — as consisting of six or more major plates which converge, move apart and slide past each other. The theory assumes that each plate behaves as a rigid unit, deforming only at its edges. Some of the Earth's most dynamic features such as volcanism and earthquakes occur at the boundaries of the plates.

Plate tectonics did not develop as an isolated event, but descended directly from earlier ideas. The most important of these was continental drift, first proposed by Alfred Wegener in 1912. Another was the concept of sea-floor spreading which, suggested in the work of Arthur Holmes, then at Edinburgh, in 1944 and Harry Hess at Princeton University in 1962, was substantiated in 1963 by Fred Vine and Drummond Matthews at Cambridge University. In a remarkable paper in 1965. J. Tuzo Wilson, a Canadian, was the first to combine the hypotheses of continental drift and sea-floor spreading into a single global concept of mobile belts and rigid plates (*Nature*, vol. 207, p. 343). Then early in 1967, at a meeting of the American Geophysical Union, scientists from Lamont-Doherty Geological Observatory provided documentary evidence for sea-floor spreading at the world's mid-ocean ridge system and for underthrusting — where one block of crust dips beneath another — at deep-sea trenches. Suddenly, it seemed a simple order had been found in the large amounts of marine geophysical and seismological data collected after the end of the Second World War. The excitement generated by these discoveries did not quickly subside. In 1967, British scientists Dan

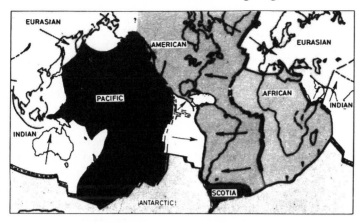

Figure 26.1   *According to plate tectonics, the outer layer of the Earth – the lithosphere – consists of a number of rigid plates which move relative to one another. New crust forms at the mid-ocean spreading centres where plates diverge; it is destroyed at boundaries where plates converge*

McKenzie and Robert Parker, and in 1968, the American W. Jason Morgan and the Frenchman Xavier Le Pichon, quantitatively combined the results from marine geophysics and seismology into the theory of plate tectonics.

Usually in science, the introduction of a new concept is followed by a period of testing and studying the implications. Certainly, plate tectonics survived a number of early counterattacks by adversaries such as Arthur Meyerhoff in the US and Vladimir Beloussov in the USSR (*New Scientist,* 12 June, 1980, p. 234). But now the concept is generally accepted by the majority of Earth scientists, although it is still being tested and its implications for the Earth sciences are continuing to be examined.

Plate tectonics has, in fact, left Earth scientists with a number of unanswered questions. The most important of these are: what are the forces that drive the plates; what causes vertical movements of plates; how rigid are the plates; and can plate tectonics be applied throughout the history of the Earth? Clearly, these are difficult questions but how we address them in the future may have a profound effect not only on the validity of the plate tectonic concept, but on how we assess the Earth's diminishing resources and live with its dynamic changes.

Perhaps because of these questions large international programmes were set up in the late 1960s and early 1970s to test and examine the implications of plate tectonics. The largest of these were the Geodynamics Project which concentrated on the continents, and the International Decade of Ocean Exploration (IDOE), the Joint Oceanographic Institutions Deep Earth Sampling (JOIDES) and the Deep Sea Drilling Project (DSDP) which emphasised the oceans. Plate tectonics has therefore enjoyed large international support over the past several years.

What will happen next in the chain of events which began with the birth of Wegener 100 years ago this week? In an attempt to address this question I will select, somewhat arbitrarily, four frontiers in research in plate tectonics that now seem to be capturing the attention of Earth scientists. These are mantle convection, sedimentary basins, earthquake seismology and continental geology. Research in these areas may alter the original concept of plate tectonics or lead to a totally new view of the evolution of the Earth.

Mantle convection

What are the forces that drive the motions of the plates? Most scientists today believe that they are driven by some form of thermal convection in the Earth's mantle – the region below the crust – where temperatures and pressures are high enough for rocks to flow. The mantle is believed to be characterised by a circulatory pattern of rising and sinking convection currents, as in a pot of boiling fluid. Curiously, we know little of how convection would appear from the surface or even whether convection occurs in a thin layer or throughout the whole mantle.

Theoretical studies by Lord Rayleigh and H. Benard in the early 1900s first established the phenomenon of convection in a layer of uniform viscosity heated from beneath. They showed that the convection was characterised by a symmetrical pattern of "cells" which were similar in width and depth.

The surface of the Earth, however, consists of small plates as well as large plates, suggesting asymmetric convection cells of different width and depth. Thus, the Rayleigh–Benard model of convection may not apply to the mantle. In computer modelling to investigate the effects of changes in fluid viscosity on convection, Dan McKenzie has found that if a thin layer of low viscosity is included in the convecting layer the convection cell becomes wide

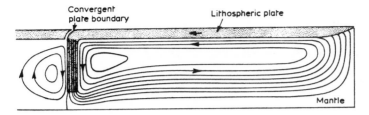

Figure 26.2 *Frank Richter's simple model for the driving mechanism of plate tectonics. The lines are streamlines showing the pattern of return flow of mantle under the moving plate on the right*

compared with its depth. Unfortunately, there is presently too little information on how the viscosity varies with depth in the mantle. Frank Richter at the University of Chicago, in another approach, has set up simple models to determine the viscosity by using geophysical observations at trenches and mid-ocean ridges to estimate the forces governing plate motions. He finds that the largest driving force is due to the sinking lithospheric slab and the largest resisting force occurs in the bottom and sides of the slab. The viscosity of the mantle is then estimated by setting up models with different lengths of sinking slab and comparing the resulting motions of the plate to observed plate motions.

These studies have not addressed the important question of why theoretical studies indicate regular size and symmetric cells while plate tectonics suggests irregular sizes and asymmetric cells. Frank Richter has suggested that two scales of convection may occur in the mantle: one scale is given by size of the plates themselves, and a second smaller scale is given by the depth of the convecting layer.

Thus, an important problem in the future is to determine the plan form of convection which exists in the Earth's mantle. Unfortunately, the physical properties of the plates themselves obscure the effects of mantle convection in observations made at the Earth's surface. However, three types of observation may help us to "see through" the plates and provide information on the convection below. These include the study of broad variations in the intensity of the Earth's gravity field which may be sensitive to the density of the convecting fluid; areas of unusual topography which may occur above rising and sinking convection currents; and geochemical anomalies in volcanic rocks which may indicate variations in the degree of mixing of material in the mantle.

Plate tectonics provides a global framework that successfully describes the horizontal motions of the plates, but what of vertical movements – perhaps revealed most dramatically in the deep sedimentary basins which have formed on continental crust during the geological past. What causes these basins to subside slowly through time?

Frontiers in Technology

The development of new technology often leads to advances in a science, and also provides the means to test a concept critically. What are the new technologies that may have important consequences for plate tectonics in the future? Two particularly promising areas are those of remote sensing of the Earth from space, and multi-channel seismic reflection profiling of the crust and upper mantle.

The most vigorous programme of remote sensing from space is presently being carried out in the US by NASA. Space techniques can determine the relative motion between the Earth's major and minor plates, the broad variations of the Earth's gravity field, and structural trends in geologically complex regions. The most promising techniques used to determine relative plate motions are very long baseline interferometry and lunar laser ranging (*New Scientist*, 16 October 1980, p. 163).

The broad variations in the Earth's gravity field have been successfully measured using radar altimeters mounted in artificial satellites. These

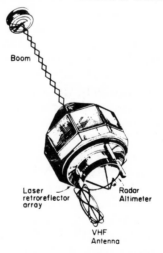

Figure 26.3   *NASA's GEOS-3 satellite*

instruments, used on SKYLAB, S-193, GEOS-3 and SEASAT-A, operate by first sending a short-pulse laser signal from the satellite to the Earth's surface and then receiving its reflection. The time elapsed between transmitting and receiving determines the distance of the satellite from the Earth's surface. By accurately calculating the orbit of the satellite it is possible to use this instrument to determine the mean height of the sea-surface, which follows variations in the strength of the gravity field over the oceans.

Information on structural trends in complex geological regions has come from use of multispectral scanners (MSS) mounted in artificial satellites. These instruments detect sunlight reflected from the Earth's surface; vegetation, soils and rocks all reflect light with different wavelengths. The LANDSAT-1 and LANDSAT-2 satellites carried an MSS system which used both the visible and infrared part of the spectrum (wavelengths 0.5 to 1.1 μm). Images based on these data are widely used in the study of continental geology, particularly in the determination of structural trends along the Eurasian/Indian plate boundary.

The application of multi-channel seismic (MCS) reflection profiling techniques to the study of the crust and mantle is presently being vigorously carried out in the US, France, Britain and West Germany. The most active programme in the continents is being carried out by the US Consortium for Continental Reflection Profiling (COCORP) which began operations in 1975. The present practice is to use long seismic arrays (6–10 km) connected to a 96-channel recording system mounted on a truck. COCORP uses five vibrating trucks as a sound source and runs profiles over geological features of 50 – 200 km in length.

The COCORP system has now been used over a range of geological features and environments in the US and Canada, and although it has not yet successfully profiled the "Moho" — the Mohorovic discontinuity separating the crust and mantle—it has revealed new information on deep layers in the Earth's crust beneath mountain ranges and fault zones.

The application of MCS profiling to the oceans was pioneered by the oil industry in its search for offshore oil and gas. MCS profiling techniques, incorporating large seismic arrays (2.4 – 3.6 km) are towed behind ships used by industry, government and academic groups to examine the deep structure of the crust and upper mantle in oceanic regions. Seismic reflection techniques have not only outlined deep sedimentary basins at continental margins but, in places, have actually continuously profiled the "Moho". These techniques seem poised to answer many important questions, in the future, on the deep structure of the crust and mantle in regions below the oceans.

Sedimentary basins

The problem of sedimentary basins is best illustrated if we consider the thickness of the sediments and the environments in which they were deposited. The sedimentary basins of North America, the Soviet Union and Europe, for example, contain several kilometres of sediments that mainly formed in non-marine or shallow marine environments. The principle of isostasy, which views that the Earth is in some form of balance with regard to its crustal thickness and density, indicates that the additional weight of the sediments is not sufficient to press down the crust and mantle beneath to form the large thicknesses of shallow water sediments observed. Other factors must be involved.

Norman Sleep of Stanford University observed in 1971 that the rate of sinking of sediments in the eastern US was initially quite rapid and then slowed down with time. The rate of sinking, he pointed out, was remarkably similar to that of a mid-ocean ridge. One possibility, therefore, was that the continental margin was initially heated up early in rifting and then sank as it cooled with time. This mechanism by itself would not necessarily produce a sedimentary basin, as the margin would be uplifted above sea-level and cooling would restore the margin only to sea-level. Sleep noted, however, that the uplifted margin would quickly be reduced to sea-level by erosion. This would produce crustal thinning at the time of initial rifting, so that subsequent cooling would drive the margin below sea-level and form a basin for sediments to infill.

This mechanism is still invoked to explain the sinking of basins at continental margins, although in a somewhat modified form. While exploring the seismic structure of the continental margin off France, Lucien Montadert, of the Institute Francais du Petrole,

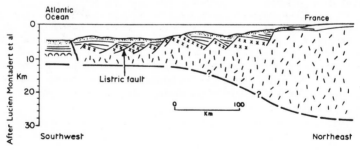

Figure 26.4 *A simplified geological cross-section of the continental margin off France shows narrow basins, believed to have formed by stretching of the crust as the Atlantic opened*

noticed that the upper part of the continental crust of margins has been fractured into a remarkable pattern of narrow sedimentary basins bounded by *listric* faults, that is, faults that "curve", being steep at the surface becoming more horizontal with depth. He suggested that the continental crust at the margin was extended at the time of rifting by up to 20 per cent. And while working on earthquake and fault patterns in western Turkey, Dan McKenzie developed a stretching model that explains the character of the sinking and the presence of listric faults at margins. This model, which differs from Sleep's model mainly in the mechanism by which crustal thinning occurs, can generally explain the large thicknesses of sediments which infilled continental margin basins during the following rifting.

The sinking of sedimentary basins in continental *interiors* is more difficult to explain as there is no obvious thermal source nearby as there is at the margins. The sinking of interior basins, such as the Michigan and Illinois basins in the US and the Pannonian basin in Hungary, is generally similar to that of the basins at margins. Stretching models have, therefore, been fitted to the data for interior basins, but this still leaves a number of problems. Most interior basins often have long geological histories and they cannot always be attributed to a single thermal event. In addition, listric faults have not yet been found beneath the Illinois and Michigan basins and even where they are observed the amount of extension indicated by the model does not clearly correlate with the amount of extension indicated on the faults.

It is important, however, to try and establish the thermal models which may explain the sinking of interior basins – not least in order to estimate how temperatures in the sediments vary with time. Many scientists now believe that whether hydrocarbons in sedimentary basins can become oil or gas strongly depends on temperature. Most of the world's oil production is from sedimentary basins, so the formation and development of basins is an important area for study in the future.

Earthquake seismology

Studies of the distribution of earthquakes at island arcs and trenches, mid-ocean ridges and transform faults (where two blocks are sliding past each other) contributed in a major way to the development of plate tectonics. Most of the world's earthquakes are located in the narrow belts which delineate the boundaries of the plates.

Three different lines of evidence – spreading rates at mid-ocean ridges, earthquake slip vectors and trends of oceanic fracture zones – give a measure of the relative motions between the plates. But these data give the movements averaged only over periods of time of up to a few million years, which is much longer than the repeat times of large earthquakes.

Plate tectonics however can aid the long-term prediction of large earthquakes. If the boundaries between plates stick, then continual plate motion would cause a build up of strain which would eventually lead to an earthquake. Of course, some of the strain along a plate boundary may be relieved by a seismic slip (that is, without earthquakes) or creep along faults. However, many "gaps" in seismicity occur along plate boundaries that have had a history of previous large earthquakes. Thus it seems likely that in these regions, strain is accumulating and there is a high "potential" for a large earthquake in the future. Lynn Sykes and colleagues at Lamont–Doherty Geological Observatory have recently constructed a global model showing regions of high seismic "potential". These include areas such as parts of the Caribbean, Taiwan and the southern Ryukyu islands, Sumatra and northern Chile.

Whereas most methods for the long-term prediction of earthquakes are based on the regularity of the seismic cycle itself, short-term predictions rely most strongly on premonitory effects such as changes in seismic velocities, ground-water geochemistry, ground height and even animal behaviour prior to an earthquake. Some earthquakes can be, and have been, predicted on the basis of these changes. However, premonitory effects appear to vary with tectonic environment and earthquake magnitude, and short-term earthquake prediction is not yet a reality although it remains an important challenge for the future.

While most of the world's earthquakes occur at plate boundaries a few per cent occur in the plate interiors. These include earthquakes of magnitude larger than about 6 in the central US, India, South Africa and Australia and a few oceanic regions remote from plate boundaries. Lynn Sykes, in 1978, carried out a detailed study of intraplate earthquakes and concluded that many of them follow pre-existing lines of weakness in the Earth's crust such as faults, mountain ranges and other tectonic boundaries.

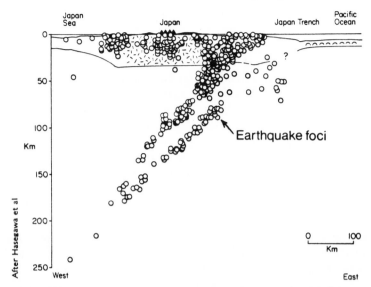

Figure 26.5   *Studies of earthquakes show a distinctly two-layered "Benioff zone" beneath Japan, believed to outline the geometry of the lithospheric slab as it sinks beneath island arcs that form behind the trenches*

Rarely, however, do earthquakes occur in very old parts of the continents or in the main ocean basins.

Intraplate earthquakes are an important problem to study in the future as plate tectonics assumes an absence of deformation within plates. Better seismic networks will be required, however, to determine the distribution and focal mechanism of these earthquakes. The study of earthquakes in regions that have experienced large vertical movements through geological time, such as mountain ranges, will be particularly interesting.

Continental geology

Much of the evidence for plate tectonics came from studies in oceans. But can plate tectonics explain the main features of the continents, such as mountain ranges? Curiously, we now know a great deal about the deep structure and evolution of the oceans but little about the continents; their geology is an important area for study.

There has, in general, been a great deal of success in using the concepts of plate tectonics to understand the geology of continents. This is quite remarkable when it is considered that plate tectonics is a concept that quantitatively describes the *present-day* motions of the plates. Geologists have found evidence for most types of plate boundary in the geological record of continental rocks. Convergent plate boundaries have appeared in an assemblage of terrains which include an accretionary wedge (that forms between a trench and island arc, as material is scraped off the downgoing slab), a magmatic arc and a fold and thrust belt. Divergent plate boundaries, on the other hand, are revealed by the continental shelf, slope and rise terrains that remain following the opening of an ocean. Various plate boundary models have been proposed to explain the late Precambrian (about 700 million years) to recent history of the Cordilleran system in western North America (the Rocky Mountains) and the late Precambrian to late Palaeozoic (about 700 to 200 million years ago) history of the Appalachian and Caledonian mountain ranges in eastern North America and Britain. These models all contain basic similarities although they differ considerably in details such as the role of strike-slip faulting, spreading behind island arcs, and the orientation of the sinking lithospheric slab.

But can plate tectonics be applied to terrains older than about 700 million years? The Archaean (about 2500 to 3500 million years ago), for example, contains only two main types of terrain: high-grade metamorphic (altered) terrains and low-grade greenstone (altered basalt) belts. So far no simple plate tectonic models have explained the high-grade terrains. The greenstone belts, on the other hand, have been interpreted as oceanic crust formed behind island arcs, so the tectonic setting of these belts would be similar to the small ocean basins in the western Pacific, such as the Mariana and Le Havre troughs. However, because of the lack of evidence of other types of plate boundary, it is still not clear whether plate tectonics, as we know it today, existed in the Archaean.

The question of whether plate tectonics can be applied to the geological record has broad implications for the Earth sciences, particularly the fields of palaeoclimatology, palaeogeography and palaeontology.

Fred Ziegler at the University of Chicago has used palaeomagnetic and palaeoclimatic data to reconstruct the positions of the continents through the geological past.

Palaeomagnetic data provide only general information on the relative positions of continents through time; but when these data are combined with climatic indicators (such as evaporites, tillites and desert sands) the former position of the continents can be reconstructed with some confidence. We have yet to discover whether the positions of the continents determined in these studies, which assume that the climatic realms of the past are similar to those of today, are consistent with plate tectonic reconstructions.

Perhaps one of the most exciting implications of plate tectonics is in palaeontology. Jim Valentine, at the Woods Hole Oceanographic Institution in Massachusetts, and colleagues suggested in the early 1970s that plate dispersal might be related to the diversification of animals. With the recent improvements in determining the relative positions of the continents through time it should now be possible to use the fossil record to quantify this relationship. Other factors, such as changes in sea level through time, may also affect faunal diversification. In the years to come the implications of these studies, and of plate tectonics, may extend to the evolution of life itself.

6 November, 1980

PART FOUR

The New Insights

27

Buffer plates: where continents collide

CONSTANTIN ROMAN

Sea-floor spreading and the resulting movement of continents have been combined into a major theory of plate tectonics in which large sections of the Earth's crust behave as rigid entities. Where these meet head-on, folded mountain chains compensate for crustal shortening. But, it now appears, the violent compression may have to be accommodated by still other means.

The theory of rigid, lithospheric plates, generally known as plate, or global, tectonics, has for the past few years attracted the attention of a great many Earth scientists. It has offered an elegant, simple model of the Earth, capable of correlating over wide areas the mechanism of earthquakes, volcanoes, the formation of island arcs, mountain belts and the opening or closing of oceans. More and more, however, geologists and geophysicits are turning their attention to the nature of the different kinds of boundaries between the major plates. Here I want to discuss a type of boundary which occurs across a continental crust.

A basic concept of plate tectonics is the definition of a plate. Usually it is regarded as a piece of lithosphere, comprising oceanic and/or continental crust, delineated by a streamline of earthquakes produced along oceanic ridges, and by trenches, major crustal faults and the remnants of the ocean floor known as "ophiolitic belts". Therefore, any such plate, marked out by an active seismic belt, is not conventionally supposed to contain within its boundries any earthquakes – it should be seismically rigid, or aseismic. However, such is not the case where the plate boundries cross parts of the continental crust which have been affected by recent Alpine-type mountain-building movements. Over wide areas such as the eastern Mediterranean, Turkey, Persia, Balluchistan, Tibet, Sinkiang and the western United States, there are numerous earthquakes, the occurrence of which would suggest that some plates are not rigid and consequently obey different rules. One way out of this dilemma was suggested by Dr D. P. McKenzie of Cambridge for the eastern Mediterranean and later on applied by A. A. Nowroozi of Columbia University to Persia. It involved breaking up the existing Eurasian plate into smaller pieces. This method of looking at the problem has the disadvantage that very small blocks of lithosphere do not play a relevant role on a global tectonic scale, only on a regional scale. Besides, many of these small plates, such as that devised for Turkey, are in fact hyperseismical, being covered all over by epicentres of shallow earthquakes. They contradict the very definition of rigid plates and prove that carving up the crust into smaller plates cannot solve the difficulties.

At Cambridge, I have studied thousands of seismograms obtained by the World Wide Seismic Network for earthquakes which occurred in Central Asia, behind the arc of the Himalayas. The particular interest of this region lies in the fact that it constitutes the world's widest seismic belt of shallow earthquakes, forming also the seismically most active area of the globe.

The study of the focal mechanisms producing these earthquakes should enable one to understand the movement of the Earth's crust in the regions of Tibet, Sinkiang, Mongolia and Lake Baikal. The implications of the study are far reaching; it should lead to a better understanding of the impact of the impingement of India onto Eurasia with its corollary, the rising of the Himalayas; and, by clarifying the plate borders in the area, should help to explain the existence of the world's widest seismic belt.

To unravel what was happening in this region I devised a new simple method of "screening" the earthquakes in terms of their frequency in each of a number of subdivisions of magnitude. Earthquakes are classified, on various magnitude scales. I have used the body-wave (USCGS) magnitude. I divided the earthquakes into half-magnitude classes and plotted the positions of their epicentres on a series of maps of Central Asia. I noticed that while small-magnitude earthquakes had no apparent preferential distribution, the larger-magnitude earthquakes tended to occur along certain, very well defined tectonic features. My results confirmed the earlier expectations of Le Pichon of the Centre Océanographique de Bretagne, Brest, which he expressed at a Royal Society meeting on plate tectonics held in 1971 – namely,

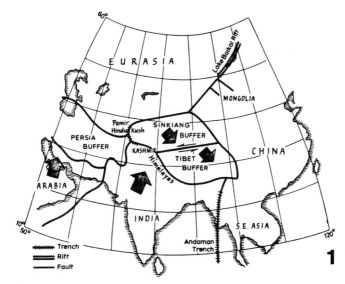

Figure 27.1 *The rigid crustal plates which India and Eurasia represent have trapped between them the two buffer plates of Sinkiang and Tibet*

that only earthquakes above a certain magnitude are relevant on a global scale.

In Central Asia, for example, it appears that behind the Himalayas there are two large hyperseismic plates, in Tibet and Sinkiang (Figure 27.1), which play the role of buffers between the rigid, aseismic plates of India and Eurasia. Actually, a buffer plate has boundaries which are outlined by major crustal faults, capable of accumulating enough strain to release large shocks of 5.5 magnitude and more. Within the boundaries of buffer plates smaller magnitude earthquakes occur throughout the crust. This definition is consistent with the structure of the Tibet and Sinkiang regions, as the newly defined buffer plates enclose tectonic entities such as the geologically recognised Tarim and Tibetan platforms (Figure 27.2).

From the purely seismic point of view I then studied the focal mechanisms of the earthquakes outlining the buffer plates of Tibet and Sinkiang and discovered a remarkable consistency in the spatial distribution of the "slip vectors", the axes of crustal compression and tension, with respect to the borders of the buffers

Figure 27.2   *Sketch map of the Sinkiang and Tibetan buffer plates, showing their major structural units. Notice the decreasing age of mountain-building movements from north to south, suggestive of continental accretion*

(Figure 27.3). The collision resulting from India's movement relative to Asia caused it to be thrust beneath the latter continental plate. The results obtained in the Department of Geophysics, Cambridge, prove that the northward impingement of India has caused most of the underthrust along the Hindu-Kush–Kashmir front of the Himalayas rather than along the lesser arcuate part of the eastern Himalayas. Here, in contrast, the thrust tends to disappear to allow for earthquakes of a different type of mechanism (strike-slip, or horizontal shearing; and at the easternmost end, dip-slip or normal faulting). The compressional axes of most of the earthquakes are contained in a horizontal plane. The focal mechanisms of a thrust and strike-slip type have compressional axes perpendicular to the limits of the buffer plates, while the mechanisms of dip-slip type have compressional axes contained within the border of the buffers.

Another important feature, capable of showing the major relative movements of plates in Central Asia are the slip vectors. The study of earthquakes shows that while the Indian plate pushes in a northerly direction, the Tibet and Sinkiang plates, which slip past each other, both move in a south-easterly direction (more

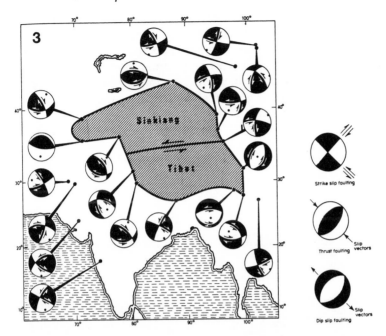

Figure 27.3 *Focal mechanisms of the earthquakes around the two buffer plates. Each mechanism is the image of the Earth in stereographic projection showing zones of compression (black quadrants) and dilatation (white quadrants).*

precisely 140° east of north). This relative movement of the Tibet and Sinkiang plates with respect to India and Eurasia is crucial in understanding the tectonic evolution of Central Asia; and therefore eventually to reconstructing the continents at the time when India was separated from Siberia by the Tethys ocean. During that particular period Tibet and Sinkiang were two continental nuclei existing somewhere between the Siberian and the Indian shields. With the gradual consumption of the Tethys oceanic floor, through the northerly migration of India, Tibet and Sinkiang were trapped between the Indian and Eurasian plates. This created in the zone of Tibet and Sinkiang two hyperseismic plates, which play the role of buffers, accommodating the major movement between India and Eurasia.

Farther along, to the north-east, the Mongolian earthquakes occur along crustal faults which are normal to the Lake Baikal rift zone, and are thus undoubtedly linked with this extensional feature. The large historical earthquakes of China, going back to the 14th century, suggest the existence of other buffer plates. They should characterise many regions affected by younger (Alpine, Mesozoic) and even Hercynian mountain building phases. They should provide the answer to the dilemma which plate tectonics faces in the eastern Mediterranean, Persia, the western United States, and elsewhere.

25 January, 1973

28

When do earthquakes occur?

JAMES BRANDER

The terrible catastrophe in Guatemala again raises the question of whether these outbursts of terrestrial wrath could be anticipated. Prediction by means of the currently fashionable phenomenon of dilatancy is likely to prove costly; while safe building programmes call for a knowledge of earthquake behaviour that has so far eluded geophysicists.

The possibility of foretelling the occurrences of natural phenomena has fascinated men for many thousands of years, and efforts to predict earthquakes have occupied as much intellectual capacity recently as did the problems of planetary orbits 400 years ago, and of sunrise 4000.

Only three countries among all those in the world that suffer from disastrous earthquakes have been able to mount the tremendous technological and logistic operation that detailed predictions require. It seems from the available evidence that China leads Japan and the United States in both the number and success of predictions. The techniques employed by the three countries differ considerably though the effects each relies on are probably simply different manifestations of the same thing, namely rock conditions in the future focal volume.

The Chinese seem to use the least technology in their programme, though they offset this to some extent by using a vast number of people. The reports suggest that almost everyone in a seismic province is expected to collect information ranging from depths of water in wells, to peculiar animal behaviour, and unexplained flaking of paintwork. There also seems to be a fairly dense local seismograph network, though sometimes a bit antiquated. There are some suggestions that they measure ground tilt and elevation changes as well, though the former of these requires some fairly sophisticated instrumentation to be accurate enough to be of any use.

The Japanese programme has concentrated on monitoring elevation changes and in establishing patterns of seismicity. One excellent example of a seismicity pattern is the so-called "doughnut" effect, which desribes how a large circular area around a future epicentre becomes seismically very quiet for a period of years before the earthquake, while an annulus beyond it experiences more than usual activity. Sometimes the annulus closes in during the time leading up to the earthquake.

The best bet for a physical basis for earthquake prediction lies in an originally very little publicised observation made by two Russian scientists. A. M. Kondratenko and L. L. Nersesov published in 1962 "Some results of the study on changes in the velocities of longitudinal and transverse waves in a focal zone", describing, in Russian, the effect which is now, since its wholesale adoption by American seismologists, known as dilatancy.

The title of the Russian paper summarises the observable seismological effects, which are simply that the velocities of the two principal seismic waves, the compressional and the shear, are decreased from normal in a volume of rock about to become the focus of an earthquake. Thus seismic waves of both types traversing the volume will take longer than other waves travelling an equivalent path length outside the future focal volume.

In fact the compressional wave velocity v_p decreases rather more, by about 15 to 20 per cent, than the shear wave velocity v_s, which decreases by only about 5 to 10 per cent. Thus the velocity, and therefore travel time, ratio for the two waves, v_p/v_s and t_s/t_p respectively, decrease from a value around 1.7 to 1.8 (for crustal rocks) to around 1.5 to 1.6. This ratio is the most convenient method of investigating dilatancy seismologically, since the travel-time ratio follows directly from the gradient of the graph of the difference between compressional and shear wave arrival times against compressional wave arrival time for a number of seismographs – the so-called Wadati plot.

But dilatancy is really a rock mechanics word used to describe the condition of a rock near to failure under stress. These changes of seismic wave velocities are only two manifestations of the overall condition.

When a rock is near to failure, the myriads of tiny imperfections and fractures in it begin to coalesce. They do so to such a degree that its volume actually increases a small amount, and the bulk modulus, an inverse measure of its compressibility, decreases. The rigidity modulus, a measure of its resistance to shear, also slightly decreases. The compressional wave velocity in a medium, be it rock, air or water, is proportional to the square root of the bulk modulus; while the shear wave velocity is proportional to the square root of the rigidity modulus. Thus the Russian seismological observations were simply a confirmation of long established laboratory results.

The onset of dilatancy can also be detected from observations of ground surface level changes, inclination changes, density changes (through gravitational effects), and magnetic changes, all of which stem from the slight volume change of the dilatant rock mass.

Limited prediction value

Where seismological observations differ from those made in the laboratory (and it is this aspect which gives dilatancy its power as a prediction tool) is that a short time before the occurrence of the earthquake the travel-time ratio increases from its anomalous value around 1.5 to around its normal value. This warning time, usually a few days, is obviously tremendously useful. But what is perhaps even more so is that there is an empirical logarithmic relationship between the total duration of the anomalous travel-time ratio and the magnitude of the resulting earthquake. Thus dilatancy provides extremely valuable prediction information.

The physical reasons for the return to normal travel-time ratios immediately before the earthquake are not completely understood, though they seem to be tied up with the pressure of water in the pores of the rock.

So we are left with the fact that dilatancy could be a very useful prediction tool. Why then haven't there been more predictions? The number of published predictions is so far very small, partly due to modesty, and also partly from fears of the social consequences, a question currently under considerable discussion in America. There has been a number of retro-predictions, situations where people have looked at travel-time ratios after the event and found them anomalous. Among these is the San Fernando earthquake of 1971, which could presumably therefore have been predicted had the significance of these anomalous travel times been appreciated at the time.

No, unfortunately, dilatancy as an earthquake predictor is only of limited application.

In order to work it requires that seismic waves propagate actually through a future focal volume and be then recorded at enough seismographs (say, four at least) to enable the drawing of a Wadati plot. Since focal volumes do not seem to be very large (though they increase rapidly with the size of the future event, and, that concerned in the 1971 San Fernando earthquake may have had a diameter as large as 80 km) a fairly dense net of local seismic stations is needed. Thus to cover even a small section of an active fault zone requires a very large array indeed. The source of seismic waves, too, is a problem since some investigators have found that artificial waves (from explosions) don't work (though why is not clear). That means that as well as the large seismograph array, suitably placed natural sources of seismic energy (earthquakes) are also necessary. Set an earthquake to catch an earthquake!

Another physical limitation of the effect is that it probably doesn't work particularly deep because the greater overburden load will tend to prevent the necessary volume increase. Indeed, the greatest depth at which dilatant volumes have been detected so far, even retrospectively, is about 15 km, half way through the Earth's crust. This depth covers all the earthquakes in California, and some of those in Japan and the Middle East, but none of those in South America.

Thus despite the tremendous academic significance or the Russian, and subsequent American, investigations into dilatancy-oriented earthquake prediction it seems to demand far too high a price in capital investment (in seismic arrays), and luck, to be a really practicable predictor for most of the world.

Long before the value of the dilatancy phenomenon was appreciated, Professor Frank Press said, in the report of the Panel on Earthquake Prediction in 1965, that to predict earthquakes measurements must be made ". . . of measured strain, measured stress, atmospheric pressure, tide potential, sea level, micro-earthquake activity, everything down to the mating habits of fish in the nearby harbour." The advent of dilatancy as a predictor has essentially only lengthened this list.

By far the most significant success in saving lives and property from the effects of earthquakes comes, and will increasingly come, from building to withstand them. If a designer can be told in advance what accelerations and displacements his building is likely to undergo during its intended lifetime, he can allow for them. To provide the necessary information, the seismologist needs to know four things about the locality: the positions of the local active faults, the local attenuation characteristics of the rocks, the sort of foundation the projected building will have, and the temporal probability distribution of earthquake occurrences. The first three items in this list are entirely local, studies from one region being of no use in another. The third, though, is a more general problem involving the hunt for patterns of crustal behaviour.

Popular enthusiasm has always been great for periodicity in the occurrences of natural, particularly devastating phenomena. An excellent example is provided by the mid 1960s earthquake scare in San Francisco which can only have arisen because the two previous disastrous earthquakes on the San Andreas fault, 1906 San Francisco and 1857 El Cajon, were seperated by 50 years. Scientific evidence available at the time about rates of movement, and the complementary nature of the two events, did not support the idea that a recurrence time as short as 50 years was at all likely. Thus it is not surprising to find that, as soon as the first earthquake lists were being produced during the last part of the 19th century, there was a great hunt for periodicities.

However, they were not to be found so easily, as the most cursory glance at the catalogue will show, so that even by 1884 astronomical periods were being invoked. In many ways this was very sensible, since we now know for sure that the Earth's shape is periodically distorted by the gravitational forces of the astronomical bodies and that this distortion is accompanied by all manner of stresses and strains. Indeed, in certain specialised environments, particularly volcanoes, there are strong indications that these tidal forces do trigger at least the smaller earthquakes.

In 1938 seismologists made one of the most complete studies along these lines, showing well-defined periodicities at all the major astronomical periods from one day to 19 years, plus a peculiar one at 42 minutes. But, alas, within months of the apparent unravelling of the seismic mystery, Harold Jeffreys pointed out that the method of analysis used was valid only if the events considered were independent. Since the earthquakes catalogue used had contained a large number of aftershocks, this condition was hardly met at all. It eventually turned out that none of the supposed peaks in the interval distribution was statistically significant.

Although, since then, there have been sporadic attempts at

discovering astronomical periods in the intervals between earthquakes – the most recent being the sidereal day, 23 hours and 56 minutes – interest among scientists has rather waned, probably because earthquake prediction by direct measurement has seemed a more realisable goal.

Vain search for non-randomness

Thus the question of what governs the temporal distribution of earthquakes (if anything) remains substantially unanswered, except possibly in the case of volcanic micro-earthquakes, as mentioned above, where Earth tide stresses do seem to play an important triggering role, and in the case of aftershocks.

Very early in the development of seismology, the Japanese scientist F. Omori found that there is a well-defined law governing the probability of an aftershock's occurring in a time interval dt at a time t after the main shock ($P(t,dt)$):

$$P(t,dt) = \frac{p.dt}{t+q}$$

where p and q are constants.

He established this probabilistic relationship on the strength of two behavioural observations of aftershock sequences, namely: that the number of aftershocks increases cumulatively with the logarithm of time after the mainshock; and that the number of aftershocks occurring within a fixed interval decreases approximately linearly with time. Both these laws have frequently been verified experimentally since, though it has often been found necessary to superpose two decay functions to describe fully an observed aftershock sequence.

Thus, once a main earthquake has occurred we have a good idea of how the aftershocks will be distributed in time, particularly if previous observations in the locality have defined the constants p and q in the probability relation. But we still do not know how the mainshocks are distributed in time.

Even while the great enthusiasm for astronomical periods still held, some doubters had been trying to fit other relationships to the observed distribution of intervals between earthquakes, notably the Poisson. If events follow a Poisson distribution it means they are randomly distributed in time, influenced neither by one another nor by any external agency. The first attempts at fitting a

Poisson process to the earthquake catalogue were not particularly successful. There were always too many empty intervals, implying some sort of temporal clustering, or that the occurrence of one earthquake predisposed the crust locally to the occurrence of another. That is only another way of looking at aftershocks, so the next attempts at fitting a Poisson process were made with obvious aftershocks removed. The fit improved greatly, though it was still not particularly significant.

In 1960 A. Ben-Menahem obtained a statistically significant fit of a negative exponential distribution to the intervals between major (greater than magnitude 7) earthquakes around the world. That did not, however, mean that the process is Poisson since other non-random processes can still give negative exponential probability distributions. Indeed, despite the statistical significance of the fit, he still had an excess of empty intervals, implying that even for these large earthquakes which were widely distributed in space and couldn't have included any aftershocks (because of the magnitude range considered) there was still some interrelation. H. Benioff came to the same conclusion from an analysis of the total global seismic energy release with time.

Recently the seismic information accumulated over the past 70 years has reached the level at which it is possible to investigate the statistical properties of the occurrence times even in fairly local areas. L. Knopoff found a statistically significant fit of a Poisson process to the earthquake catalogue from southern California for events of magnitude greater than 2.8, though only after he had scrupulously removed all possible aftershocks. Indeed, he believes that all the non-randomness found by earlier workers is attributable to the presence of aftershocks in the list. But even his results show some distortion towards an excess of empty intervals.

Investigations into the distribution of intervals between very small microearthquakes from the same fault shows that they are quite well described by the so-called gamma distribution which includes a parameter to allow for clustering within a group of events as well as a parameter determining the relationship between groups. At this level there is certainly no doubt about the randomness of the occurrence times. It seems also from these very local investigations that the degree of clustering into groups drops quite markedly as the lower magnitude limit for the sample under investigation is raised, i.e., that larger earthquakes approach more closely a truly random temporal distribution.

In summary, it seems that, except on volcanoes, neither

earthquakes nor microearthquakes occur at times governed by astronomical behaviour. Indeed, the larger earthquakes, even in a relatively limited region, are, without their aftershocks, significantly random in time. They are not completely so, though, and become progressively more interrelated as their magnitude decreases. Aftershocks as a class of earthquakes on their own show the only easily described temporal behaviour.

So we must conclude that, despite popular fatalism, earthquakes are governed by as complicated a set of rules as any phenomenon, rules that haven't yet been completely formulated but will most likely bear more relation to the physics of continua than to some mysterious *deus ex machina*.

12 February, 1976

GUATEMALA'S CATASTROPHE

In December, 1972, the capital city of Nicaragua, Managua, was 80 per cent destroyed by a moderate sized earthquake. Some 10 000 people lost their lives and the city ceased to exist as a viable centre. Now Guatemala has suffered a similar catastrophe, from a pair of earthquakes, the first somewhat larger than Nicaragua's.

The Central American isthmus is a very complicated geological zone, with the deep Middle American Trench lying just off the west coast, and three massive fractures striking hundreds of miles westwards across the ocean floor. The area is well accustomed to violent seismic and volcanic activity and, as in many cultures, indigenous building methods have evolved over the centuries which allow for these hazards. Single-storey adobe and timber houses with wooden roofs do not lead to much loss of life and are easily repaired.

However, there has been recently a tremendous increase in urban populations, forcing the inhabitants to build upwards. This they have done both by accommodating extra storeys on their traditional buildings and by using concrete. In Managua there was no construction code of safety requirements in operation before the earthquake, though some of the largest commercial buildings (principally American banks) were built according to Californian requirements. Guatemala City, with its substantial tourist industry, was rather better endowed with such buildings, which in most cases survived to some extent.

Figure 28.1 (*Above*) *part of Guatemala City 4 February, 1976 (Credit: Associated Press)*

Figure 28.2   *Before the February 1976 earthquakes much of Guatemala City, completely destroyed in 1916, consisted of single-story houses. (Credit: Michael Freeman)*

In Managua almost all the adobe buildings collapsed, a fact which accounted for most of the loss of life. Moreover, not one of the city's four hospitals, nor its Red Cross depot, nor its fire station, survived the earthquake, so that there were effectively no emergency services available after it. Guatemala has not, fortunately, been so badly disrupted and the primary problems seem to be the almost commonplace ones of shortages of drinking water, food, and medical supplies. Even so, there has been a terrible disaster with great loss of life, mostly due to appallingly inadequate construction standards.

12 February, 1976

29

The seismic rumpus in the western Pacific

Over the past few weeks we have seen two major earthquakes and continuing seismic activity in South China and the Philippines. Like other major crustal movements on the Earth these reflect adjustments to sea-floor spreading. The big shock, and calamitous tsunami which followed it, in the Philippines marks part of the line along which the Pacific plate dives down beneath the edge of Asia. In point of fact the Tangshan earthquake, which killed an estimated 100 000 Chinese, occurred by a shear movement along an east-west fault (see map) which has previously been implicated in earthquake disasters. Last week's relatively innocuous shock at Chengtu in Szechwan 800 miles away occurred along the same fault system.

To see how the Philippines fit into a related seismic picture it is necessary to look, not at the jostling of the world's six major crustal plates, but at the minor plates which, in places, hold them apart and help to accommodate their movements.

Early in 1973 *New Scientist* (see page 197) carried an article describing a new concept in plate tectonics of "buffer plates". According to the author, Constantin Roman then at Cambridge University, such non-rigid plates helped to alleviate the crustal stresses piling up in Sinkiang and Tibet between India and Asia. On the Tangshan earthquake Dr Roman commented: "A buffer plate has its boundaries delineated by major earthquakes but is peppered all over by epicentres of minor earthquakes . . .

"China could equally fall into this category acting as a buffer between Eurasia to the north, Sinkiang and Tibet to the west, India to the south, Philippines to the south east and the Pacific to the east."

Dr Roman (who supplied this map) went on: "The two Chinese plates slip past each other along a right lateral strike-slip fault

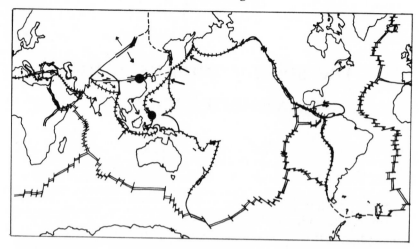

Figure 29.1 *The Earth's major and minor crustal plates showing (arrows) the directions of crustal movements around China. Full circles indicate the approximate positions of the Tangshan and Philippines epicentres. South China may serve as a "buffer" plate*

accommodating movement resulting from the spreading axis of the Baikal rift to the north west, the consumption of oceanic floor of the Indian, Philippine and Pacific plates, and the rotation of the Sinkiang and Tibetan buffer plates."

IS THIS THE BASIS OF CHINA'S PREDICTION METHODS?

It is likely that any eventually successful method of predicting earthquakes will have to rely on a selection of physical criteria. Already seismologists have discovered a number of crustal changes which, on a fairly long-term basis, may herald an impending shock. The Chinese researchers apparently employed many of these in their reported succesful prediction of the major earthquake which struck Haicheng in Liaoning Province in February last year (see *New Scientist*, vol. 69, p. 391).

What remains particularly difficult is to say just where and when a shock will occur. The Chinese allegedly looked for so-called foreshocks as the basis of their short-term prediction.

(Accounts of unusual animal behaviour anticipating Chinese earthquakes probably relate to foreshock activity.) Now two seismologists from MIT have followed up their lead by a study of the foreshocks which have accompanied major earthquakes both according to past Chinese records, and others from elsewhere of more recent date (*Nature*, vol. 262, p. 677). They conclude that a "premonitory slip" of this kind may be a feature of all major earthquakes.

It has been common knowledge among Earth scientists that major earthquakes are usually followed by a series of *after*-shocks as the disrupted crust of the Earth adjusts itself to its new configuration. Foreshocks have been less carefully documented. Lucile Jones and Peter Molnar searched Chinese records dating from 1900 to 1949 for reports of seismic events preceding all earthquakes bigger than magnitude 7 on the Richter scale. The lists are far from complete – partly owing to a lack in the earlier years of sufficiently sensitive instrumentation to spot these minor tremors. Nevertheless, they found that 21 per cent of the earthquakes were anticipated by foreshocks.

When they turned to the world catalogue for 1950 to 1973, compiled by the US National Oceanic and Atmospheric Administration, they found a much higher correlation – 44 per cent of the larger earthquakes leaked their energy beforehand in this way. Even this figure, they reckon, is too low because of the way the data were accumulated.

Plotting a histogram to show how soon the foreshocks occurred before the main earthquake, the two workers revealed that 43 per cent happened within 24 hours of the big shock, and 82 per cent within two weeks.

With better instrumentation such foreshocks, they contend, might help considerably in the prediction of many, though not necessarily all, earthquakes. One difficulty is that of distinguishing small foreshocks from the seismic background level.

"MONITOR", *26 August, 1976*

30

Turkey's earthquake

JAMES BRANDER

Last week's major earthquake in Turkey occurred, like others before it, along the Anatolian Fault Zone, a feature comparable in size to the San Andreas Fault in California. Lacking the resources of the latter region, however, prediction in Turkey may have to rely on spotting likely "gaps" where seismic stresses may be building up dangerously.

Last Thursday came in the first reports of what has this year become an all-too-familiar event, an earthquake disaster. This time it was high in the mountains of Eastern Turkey, near Lake Van, where the winter weather conditions are making the plight of the survivors even worse than usual.

This earthquake occurred on one of the few known major fault zones which had till then not contributed to this year's run of disasters, the Anatolian Fault. It is one of the Earth's largest structural boundaries, which divides, according to the plate tectonic theory, the Turkish plate from the Black Sea plate, and is marked by a 1000-km-long southward-facing arc of closely spaced earthquake locations running from the Aegean coast near Istanbul across to Lake Van. On the ground numerous fault scarps testify to the activity of the zone which has produced 31 major earthquakes since 1909.

Seen in a world context the Turkish earthquake zone is a section of one of the two great global seismic belts, the Mediterranean and trans-Asiatic zone (the other loops most of the way around the Pacific). This belt represents the tectonic activity of the last phases of uplift of the great Alpine–Himalayan mountain belt, itself a manifestation of the north–south collision of Africa and India with Europe and Asia. The lands of the European parts of this global belt most frequently blighted by earthquakes are Turkey,

Greece, and Italy. Other parts of it are not entirely safe from such disasters, however. As recently as 1954, for instance, there was a strong earthquake in Spain which originated beneath the southern slopes of the Sierra Nevada, near the western end of the belt.

All investigations of the sense of movement along the Anatolian Fault show it to be mainly a right-lateral strike slip, which is to say that the northern side of the fault moves eastward with respect to the southern side. The sense of motion, the length, and the depth of the fault (determined from the seismic records) all show the Anatolian Fault to be very similar structurally to the more infamous San Andreas Fault which threatens the cities of San Francisco and Los Angeles in California.

But in terms of devastating earthquakes this century, the Anatolian Fault seems to be the more active, though, since over most of its length it lies in very rugged mountainous terrain difficult of access even during the summer, it has received nothing like its fair share of scientific attention.

The problem of prediction

There is no question of there being sufficient financial or academic resources in the area to mount an earthquake prediction programme on the lines of that currently being developed in the US, so that precise forewarning of disasters like the present one is ruled out for the foreseeable future. But by looking carefully at the effects of the last 70 years' earthquakes, and their distribution along the fault, it is tempting to draw certain conclusions about the *location* of future activity. Unfortunately it doesn't really look feasible to make any estimates of the *times* of occurrence of this activity.

Figure 30.1 is a sketch map of the region with the Anatolian Fault Zone drawn as a dashed line marking the observed surface breaks resulting from this century's earthquakes. The amount of right lateral movement observed is typically around 2 metres.

Since 1967, the year of the last major earthquake on the Anatolian Fault, in the west, there had been two quite noticeable "gaps" in the line of epicentres and observed fault breakage, one about 300 kilometres west of Lake Van, and one due south of Istanbul. From the reports of the destruction caused by last week's disaster it seems as though this line of observed breakage should now be extended east of Lake Van, possibly leaving a third gap

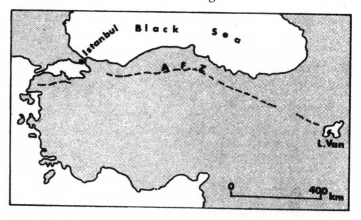

Figure 30.1 *Sketch map of the region of the Anatolian Fault Zone in Turkey. Lake Van in the east is where the recent earthquake occurred. AFZ denotes the Anatolian Fault Zone: the dashed line shows where there has been observable ground breakage (involving fault displacements of usually around two metres) during the 20th century. Note the seismic "gaps"*

immediately to the west of the lake. Could it be that these gaps are to be the sites of earthquakes in the near future since, after all, there must be quite considerable concentrations of stresses inside them? The western gap is perhaps not so much at risk as the eastern ones since three major stress-relieving earthquakes occurred on the western end of the Anatolian Fault during the second half of the last century, including the great earthquake of Istanbul of 1894.

But apart from these gaps there is also the question of the ends of the Anatolian Fault. To the west it seems to link with the complex Balkan activity, responsible for disasters like that of Skopje in 1963, but no-one really knows what happens at the eastern end. It seems probable that in some way the Anatolian Fault is linked to the faults of both northern and western Iran where earthquakes often occur in the highlands of the Alborz and Zagros mountains. Iranian Azerbaidjan, which lies between the eastern end of the Anatolian Fault and the Alborz mountains, is known historically for its seismic activity, though it has been quiet recently. Perhaps here too are other seismic gaps waiting to be filled in by devastating earthquakes.

Whether the seismic gap idea has any real value as a reliable prediction tool I wouldn't like to say, though it has been invoked before with some success. However, it is only by trying peculiar approaches like this that any chance of averting repeats of last week's disaster will ever be found for the millions of people who live on active seismic belts outside highly instrumented California and some parts of the USSR.

2 December, 1976

31

A fiery fate for Heimaey

STEPHEN SELF AND STEPHEN SPARKS

The recent volcanic eruption on the island of Heimaey is an economic disaster for Iceland. Here two geologists give an on-the-spot description of the new volcano.

At about 2 a.m., Tuesday, 23 January, 1973 a volcanic eruption began on the eastern side of the island of Heimaey, one of the Westmann Islands to the south of Iceland. A 1.5 km fissure running north–south had formed to the east of the 7000-year-old volcanic cone of Helgafell. Along the fissure there were between 15 and 20 active vents and several lava streams flowed east towards the sea. About 13 hours before the eruption several seismic stations in Southern Iceland recorded a swarm of 120–130 earthquakes of between magnitudes 2 and 3 (Richter's scale) but the epicentre was not located.

Iceland sits astride the Mid-Atlantic Ridge, a major submarine mountain range running the length of the Atlantic. The ridge is seismically active and there are records of many recent volcanic eruptions along its length – such as a Tristan da Cunha (1961) and at Fayal in the Azores (1957). The ridge runs through the centre of Iceland in a relatively narrow zone in which all the eruptions in Iceland over the last decade have occurred. Heimaey lies in this relatively new volcanic zone (probably 2–3 million years old) and is part of a remarkable sequence of recent activity in Iceland. The Mid-Atlantic Ridge is thought to be a major zone of tension in the Earth's surface along which new crust is being generated.

On the second day of the Heimaey eruption the fissure shortened to about 250 metres and only three vents were active. By Thursday the eruption centralised to a single vent 500 metres from the northern end of the fissure. Since then activity has been confined to this vent, building up a large "scoria" cone out of

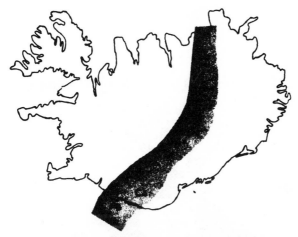

Figure 31.1 *The main zone (stippled) of volcanic activity in Iceland during the past century coincides with the axis of the Mid-Atlantic Ridge. Triangles indicate sites of known activity: A, Surtsey (1963–67); B, Heimaey (1973); C, Katla (1918); D, Hekla (1947 and 1970); E, Grimsvötn (1934 and others); F, Askja (1926 and 1961)*

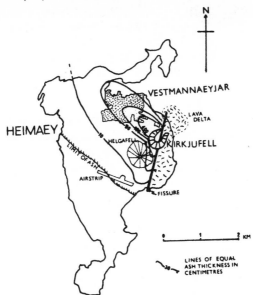

Figure 31.2 *Map showing ash engulfment (in centimetres) of the township of Vestmannaeyjar (dotted). The old cone of Helgafell is to the south-west and the main fissure is to the east*

Figure 31.3 *Kirkjufell in action, 1 February 1973, view from the sea looking south. A, westerly wind; B, turbulent ash and steam, often in a mushroom-shaped cloud; C, ejected ballistic bombs; D, lava fountain; E, 33° depositional slope of loose scoria and bombs; F, small deep orifice in cone; G, lava channel exit from cone; H, fine wind-blown ash falling from cloud; I, lava encroaching the sea with dense steam clouds above; J, old coast line; K, buried house and telegraph lines; L, bulldozers removing ash give some idea of the scale; M, old, now ash-covered, Helgafell volcano*

ejected basalt fragments, now 185 metres above sea level and named Kirkjufell by the islanders. Kirkjufell is growing through continuous fire-fountaining, which ejects lava fragments over the rim of the cone. Individual incandescent lava fountains rise between 200 and 300 metres above the base of the cone. Large flares of burning gases occasionally leap 50 to 100 metres above the crater rim. The eruption column of steam, gas and dust reaches a height of between 6 and 10 km. A lava flow is heading slowly into the sea to the east of Kirkjufell causing a breach in the cone.

The lava was moving at only 10 cm a minute on Wednesday, 31 January, and had created a square kilometre of new land in a lava delta.

Ash from the eruption has been dispersing in two distinct ways essentially dependent on the strength of the prevailing wind. In calm weather, with low wind velocities, lava fragments are ejected explosively and maintain a ballistic trajectory throughout flight. Individual fragments have a short range and pile up to build up the Kirkjufell scoria cone. Large fragments are thrown furthest because they are less affected by air resistance. Usually the red-hot lava bombs land on the cone surface and roll down the slope of 32° to 34° – an angle which represents the equilibrium slope of the material. Occasionally a particularly powerful explosion or an angled blast can throw metre-sized bombs 500 metres from the base of the cone.

By contrast, in strong winds glowing fragments are carried considerable distances (6 cm fragments have travelled 4 km from the vent). Now small fragments are transported further than large ones and the ash spreads over a much wider area. In moderate winds dispersal is a combination of ballistic ejection of large lava bombs and wind dispersal of finer material.

Figure 31.4 *250 to 300-metre fire fountain seen at night 1 kilometre away in the eastern quarter of Vestmannaeyjar*

Figure 31.5   *View of the northern part of the Kirkjufell cone showing ash-covered lava field and buried telegraph poles. Fine grained material from ash laden cloud is being blown north-eastwards*

Strong easterly winds during the weekend 26–29 January carried hot ash over the town of Heimaey causing most of the damage and threatening the town with burial. Near the cone on the east side of the town ash has accumulated so far to 4 metres, burying some houses completely. Even on the west side, furthest from the vent, 20 cm of ash has formed. So far fire due to rapid accumulation of hot ash has destroyed 39 houses. Many others are badly damaged, some having collapsed under the weight of ash.

The fire-fountaining of Kirkjufell is typical Strombolian activity producing only coarse material. Very little material of less than a millimetre in diameter is being produced. Lava bombs break up explosively during flight and can disintegrate further on landing to produce finer material. The mean of the size distribution of deposited fragments is about 1–3 cm. Lava bombs up to a metre diameter are plentiful within 500 metres of the vent.

Icelandic geologists estimate that so far 40 million cubic metres of lava and 10 million cubic metres of pyroclastic scoria have been produced (this represents about 2.5 per cent of the total volume of the Surtsey eruption). The material contains virtually no large crystals and chemically is a rather viscous basalt known as Hawaiite. The composition is similar to the Surtsey lavas which erupted in 1963, 10 km to the south of Heimaey.

Economically, the eruption is a major disaster for Iceland. Vestmannaeyjar town has a population of 5273 and as such is the fifth largest in Iceland. It plays an important role in the fishing industry and the value of export production of fisheries products for 1972 is estimated to be £5.9 million, 11.4 per cent of the country's total production. Until the eruption ceases reliable estimates of the damage cannot be made, but the cost to Iceland is likely to be considerable.

8 February, 1973

32

The volcano that won't lie down

MARGARET EVANS

Mount St Helens ended its 123 years of slumber on 20 March, 1980. Seismic rumblings developed within a few days into a major earthquake, and steam erupted from the mountain on 27 March. So far there have been four eruptions, and no one knows how much is still to come. In this article Margaret Evans describes events to date; in Chapter 33, Professor Basil Booth explains what lies behind the greatest of the eruptions.

At approximately 8.45 p.m., on 12 June 1980, Mount St Helens erupted in its fourth major explosion in the present bout of activity. The latest eruption, heard 215 km away, was accompanied by very intense "harmonic" tremors that began at approximately 9.10 p.m. and continued until around 11.30 p.m. Then the tremors subsided considerably, although they continued to be detected until about 9 a.m. the following day. The volcano blasted steam and ash in a plume as high as 15 000 metres, which an Eastern Airline pilot first spotted at 8.45 p.m. Initially the blast consisted predominantly of ash, but when the harmonic tremors began to subside the volcano expelled mainly steam, which formed a thick dense cloud above the volcano.

Material from the latest explosion appears to be slightly different from that emitted earlier. The material is darker in colour and of a different chemical composition, which leads scientists to believe that it is coming from a greater depth in the volcano than before. All the material spewed from the volcano is analysed under a polarising microscope to determine its precise mineral content.

The eruption on 12 June was not as devastating as the original major explosion on 18 May, having less substantial environmental effects. A large amount of rain falling immediately after the eruption helped confine the fallout of ash and minimise the amount of dust.

The ash from this and the three previous fallouts is a major concern. Its silicate component resembles ground glass and, when airborne, tiny particles can literally bring machinery to a grinding halt. Car engines have been immobilised, and timber harvesting of felled trees has become a Herculean task because the ash on the logs dulls a chainsaw in moments. All agricultural machinery is vulnerable to engine breakdown, as are the turbine blade at hydroelectric plants that receive waterborne ash. Profits for agriculturalists will come from the volcano, but they must wait with Job's patience for the weather to break down ash in the soil and so release fertilising minerals.

As with the previous eruptions, the blast on 12 June brought more misery to those living in the fallout zone. By Monday in Portland, Oregon – 70 km to the south – motorists were using their headlights at noon and residents were wearing air filtration masks as they washed dust from sidewalks and driveways. A cloud of volcanic ash hovered over the city for the third consecutive day.

The latest observations in the crater of the volcano indicate the formation of a dome, which geologists think may be created as a very viscous form of lava exudes and slowly builds up on the crater's floor. The dome was first observed on 15 June; it could however have formed earlier, remaining undetected because of poor visibility. By the morning of 18 June, the dome, originally observed to be 220 m in circumference and 40 m high, had doubled in height to 80 m. From the air, observers have noticed that it glows in the dark. "The glow," says Pete Towley, a geologist with the US Geological Survey, "is emanating from below, through cracks in the crust of solidified lava." The dome is growing as its upper surface cools and cracks while more hot material oozes upwards. It is surrounded by a shallow moat of water that continues to trickle down from the inner walls of the crater.

Scientists expect the dome eventually to create a cone similar to the volcano's peak before the eruption on 18 May ripped 400 m off the top of the mountain. Until then, explains Towley, it is likely that the dome will periodically collapse to produce more eruptions of ash and flows of hot gas and ash. Before Mount St Helens quietens down there may even be a lava spill-out – the mountain is just as dangerous as ever. The dome may explode with a force capable of killing anything near the crater.

Very little is known about the volcanic patterns of the Cascade Range, the mountain chain along the north-west coast through the US and Canada. Therefore everything that is happening to Mount St Helens is "a classroom" experience for geologists and scientists scrambling to gather as much data as they can with seismic recording instruments, tiltmeters and water level gauges. Many of the data are a long way from being analysed.

A characteristic of Mount St Helens that is similar to other volcanoes, such as those on the Hawaiian Islands, is the presence of harmonic tremors accompanying volcanic activity. Harmonic tremors are regular rhythmic Earth movements that result from the movement of subterranean magma. According to Towley tremors cannot be recorded on the Richter scale in the same way that an ordinary earthquake, releasing sudden short bursts of energy, is recorded. Harmonic tremors are, however, recorded on seismic instruments and may either herald or accompany volcanic eruptions, as they have in each of the four major blasts of Mount St Helens.

Will Mount Baker be the next?

All scientists appear to agree that there is a volcanic pattern existing in the Cascade Range. One of the theories is that if one volcano awakens from dormancy, others in the mountain chain will follow likewise. Records indicate that when Mount St Helens has erupted in the past, other volcanoes have become active. At the present time much attention is being paid to Mount Baker, just south of the border between Canada and the US. Mount Baker and Mount St Helens erupted more or less simultaneously in 1843, 1854 and 1857/58. In 1975 Mount Baker, the most thermally active volcano in the Cascade Range, showed increasing activity in its fumeroles – vents of steam through volcanic pressure valves. Emmissions of steam then led to the closure of the Mount Baker National Park. The mountain has continued to steam periodically since then, as has Rainier, another mountain that is being closely observed.

Scientists at least agree that the coastal volcanoes are all connected in their subsurface roots, some 200 km down. According to Richard Armstrong, a geologist at the University of British Columbia, "we are in the middle of a period of vigorous volcanic mounting building". He claims that 50 million years ago there was an enormous amount of volcanic activity in British Columbia, which continued for millions of years. As recently as 2000 years ago huge eruptions of ash occurred, creating regions of absolute desolation with mudflows, river blockings, and ashfalls – similar to what is happening in the Mount St Helens region now.

26 June, 1980

33

. . . and how the mountain exploded

BASIL BOOTH

Mount St Helens volcano, Washington State, entered a violently eruptive phase on 18 May, after 52 days of mildly explosive activity.

The new eruption began at 8.31 a.m., when an immense detonation, that was heard as far away as 320 km, blew out the north flank of the volcano and released searing hot pyroclastic surges (*nuées ardentes* or glowing avalanches) that cut a swath

Figure 33.1 *Mount St. Helens as seen from Portland for the past 123 years. Previously rising gently to 2950m, the mountain is now 400m shorter, and has a flat top. Mount Adams is in the background (Credit: Wide World Photos)*

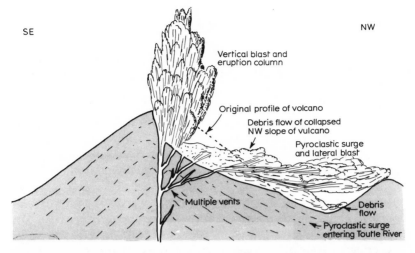

SE

NW

Vertical blast and
eruption column

Original profile of volcano

Debris flow of collapsed
NW slope of vulcano

Pyroclastic surge
and lateral blast

Multiple vents

Debris
flow

Pyroclastic surge
entering Toutle River

Figure 33.2   *The moment of eruption: 8.31 am, 18 May.
High gas pressures inside the volcano combine with a 5
Richter earthquake to break up the mountain's north flank,
and the resulting avalanche slides into the Toutle river valley.
Lateral and radial fissures allow blasts of gas-fluidised
magma, at 800°C, to form pyroclastic surges to the north.
The debris mixes with the water in the Toutle to form
volcanic mud flows or lahars, which bulldozed the river's
waters into a destructive wave that swept away camps and
bridges and buried houses*

over 20 km wide and 8 km long through the pine forests which
surround the volcano. The hurricane-like winds which accompanied the blast felled trees over a wide area and left animals in a
state of shock. Within minutes a dense plume of volcanic ash rose
above the summit to a height of 14 500 m, creating violent
lightning displays that caused widespread forest fires although the
ash fall which followed extinguished most of them.

The eruption affected a large area of the western states, with
acid ash (*p*H 5.2–5.5) accumulating to a depth of 1.5 cm at a
distance of 800 km. East Washington state was centimetres deep
in ash which fell so thickly that automatic street lighting switched
on. Some schools and businesses were closed down in Washington
and Idaho, while roads in Montana were sealed off because of the
poor visibility caused by the falling ash. Ash also fell in Wyoming
on 19 May, where children were kept indoors because of the effect

of the tiny glass shards in the ash on the respiratory system. The ash cloud moved east over Dakota toward the New England states and a slight fall was reported from Denver on 19 May, where it caused a burning sensation in the eyes. By 20 May, the ash had reached Manitoba and Ontario, 2000 km from the volcano, where it reduced visibility down to 5 km.

A routine observation flight provided valuable information on events during the opening stage of the eruption. Blocks of ice from the summit glacier and rocks were seen to fall into the crater. Next, the north flank of the cone began to vibrate and a huge bulge quickly developed. This inflated area then foundered and slid to the north, relieving pressure inside the volcano. Once relieved of this lithostatic load, the gases expanded with explosive force and blew out the fractured and weaker parts of the cone. The explosion devastated an area in excess of 380 sq. km and some blast effects extended 22 km to the north-west and 19 km to the north. The lateral blasts through the collapse fissures formed

Figure 33.3   *Mid-April: half way between the first seismic rumblings and the major explosion of 18 May. Arrows indicate the growing radial cracks that later formed the deep fissures bounding the collapse sector (Credit: B. Booth)*

Figure 33.4 *The major collapse, now well established, rapidly avalanches into the Toutle river valley. Pyroclastic surges (S) form on major radial fissures to the north and west. The vertical eruption plume is still rising rapidly (Credit: Associated Press/VERN)*

pyroclastic surges which travelled across intervening ridges as though they had not been there. Near the volcano, trees were stripped completely from the surface and further away were felled parallel to the blast.

The collapse debris flowed slowly at first but melting snow and ice blocks in the flow combined with water in the Toutle River to generate lahars (volcanic mudflows), whose velocity increased to as much as 80 km/h. The lahars bulldozed the waters in the Toutle River with a massive wave 10 m high which plunged downstream for over 19 km causing tremendous damage – sweeping away logging camps, destroying bridges, burying houses, and producing a turbulent flood packed solid with trees torn from the surrounding valley sides.

The deposit left by the debris flow had an average thickness of 66 m in the upper reaches of the Toutle River, increasing to 130 m in places. Large ice blocks were carried along by the lahars whose temperatures initially were close to 100° C in places – the result of pyroclastic surge material becoming admixed with the debris flow and water.

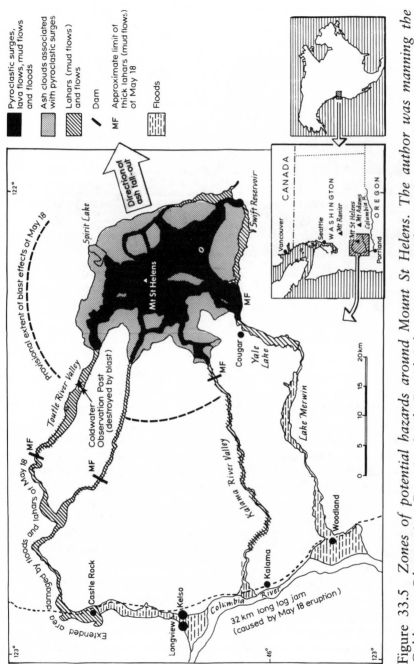

Figure 33.5 Zones of potential hazards around Mount St Helens. The author was manning the Coldwater Observation Post in April. David Johnson was at the post on 18 May, when it was obliterated. (Map is after Crandel and Mullineaux, 1978)

It appears that water was not essential during the early stages of the debris flow; gas and steam provided the fluid medium necessary for transport. Further down the valley water became essential to movement of the debris and ultimately was instrumental in generating the destructive lahars.

The flows were driven by sector collapse of the volcano, the lateral blast (producing pyroclastic surges), and by the fall-out from the eruption column that followed. The eruption column ascended eventually to 27 000 m, first driven by the high muzzle velocity (the speed of which it was hurled from the volcano, estimated at over 200 m per second) and secondly by convection rise. The next eruption, which took place six hours later, was the most voluminous and produced dense ash clouds which were driven by jet streams across the entire width of the US.

The collapse of the northern slope, together with the explosion, produced a north-facing amphitheatre whose base sloped northwards from 2800 m at the conduit to 1470 m. The almost vertical

Figure 33.6 *Well developed pyroclastic surges are moving rapidly away from the collapse. The huge amphitheatre is now clearly visible as the debris flow gathers momentum. The vertical eruption plume is still ascending rapidly with no sign of collapse (Credit: Associated Press/VERN)*

Figure 33.7 *After four minutes the volcano is totally obscured by an ash cloud 32km across. To the right the surge has begun to spill over intervening ridges. The vertical eruption plume (P) is partly hidden by ash elutriated from the surges (Credit: Associated Press/VERN)*

southern wall of the amphitheatre was initially between 500 and 560 m high.

The volume of this northerly collapse, plus that of the ash plume, has been estimated at 3.7 cu. km; most of this volume, however, is accounted for by the debris flow that followed the collapse. The value given here is provisional; more accurate determinations will require several months of field work after the area has been declared safe.

No hope now exists for the 88 persons missing which, together with the number of dead recovered, brings the death toll to over 100.

Spirit Lake, which had been filled by the debris flow and was thought to be holding a second potentially dangerous lahar, later began to drain. However, further mudflows, smaller than the first, may form when streams start to cut down through the new lahar deposits. There is also the possibility of a second blast from a lava dome that is now growing in the throat of the volcano. This could initiate further pyroclastic surges and lahars and therefore makes the north side of the volcano a high-risk hazard area.

The course and volume of the destructive lahars followed the predictions made by the US Geological Survey. However, what they did not anticipate was the blast that removed a large portion of the northern face of the volcano.

Several important conclusions may be drawn from the recent explosive eruption of Mount St Helens.

First, serial photographs of the blast demonstrate conclusively that the pyroclastic surges were not produced by column collapse, as many have supposed. The surges clearly were vented through radial fissures that developed as the cone collapsed. Later fallback from the eruption cloud (bulk subsidence or column collapse) produced only minor surges on the steeper slopes of the volcano (ash avalanches). Consequently, fissure vented surges may be more important in producing pyroclastic flows and surges during large volcanic eruptions than is generally believed.

Secondly, small amphitheatre-shaped collapse depressions on volcanoes undoubtedly can be produced by mechanisms similar to those described here. The gigantic eruption of Bezymianni, Kamchatka (USSR), in 1956, was similar in many respects to that of Mount St Helens. It produced a devastating blast and pyroclastic flows and surges which travelled down valleys opposite the collapsed sector of the volcano and later developed into lahars. The corollary is that large sector collapse depressions on many of the world's volcanoes may have been initiated by such mechanisms. If so, this may help explain some of the anomalous pyroclastic rocks (ignimbrites) that originate from the welding together of hot debris from *nuées ardentes* and with which sector collapse depressions are associated.

Lastly, because the initiation of laterally directed blasts is poorly understood, we cannot accurately define hazard zones around volcanoes. Clearly the recent events on Mount St Helens indicate that more research is needed to determine the failure characteristics of volcanoes.

26 June, 1980

34

Krakatoa: the decapitation of a volcano

ALAN WOOLLEY AND CLIVE BISHOP

One hundred years ago the world was shaken by the greatest eruption ever recorded by man.

Of all the volcanic eruptions in historic times, three have achieved the status of legend. The first was that momentous rejuvenation of Vesuvius in AD 79 that took the Romans completely by surprise, as it had not at that time erupted within living memory. Pompeii and Herculaneum were overwhelmed, Pliny the elder suffered a fatal heart attack in the excitement, and his nephew recorded the events in a manner which can still excite layman and volcanologist alike. The second is the eruption of Mont Pelée on the island of Martinique in 1902, which destroyed the town of St Pierre at the foot of the volcano – all but two of the 30 000 inhabitants being killed by a *nuée ardente* (glowing avalanche) within a few minutes. That eruption was described, though not witnessed, by the French geologist A. Lacroix, whose voluminous and superbly illustrated account is a worthy monument to the catastrophe.

The third eruption, and the subject of this account, is the eruption of Krakatoa in 1883, which, in many respects, could be considered as the most notorious of all. Hollywood has celebrated this eruption, as only Hollywood knows how, in the film *Krakatoa East of Java*, but it does not need exaggeration because it is surely dramatic and horrifying enough for the most jaded palate. Thirty-six thousand people lost their lives, most of them drowned, in the huge tidal waves (tsunami) which were a consequence of the eruption, but others were burned or poisoned by showers of hot ash. There were hundreds of explosions the largest of which was heard almost 5000 km away, and there were dramatic and enduring meteorological effects that lasted several years and which were observed all over the world.

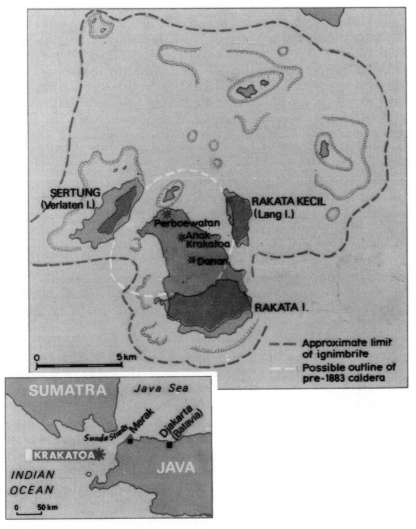

Figure 34.1   *Map of Krakatoa showing outline of the islands before (shaded) and after (solid line) the 1883 eruption. Banks of pumice formed after the eruption are shown stippled. (Based on Figure 1 in the Royal Society Report of 1888, and Self and Rampino, 1981.) Inset map shows location of Krakatoa*

But the events alone would probably not have ensured that we recall what happened a century ago had it not been that, like the eruptions of Vesuvius and Pelée, the events of August 1883 were accurately recorded for posterity. One of the two classic accounts is that by R. D. M. Verbeek, a Dutch geologist who lived in western Java and who visited the devastated areas a few days after the eruptions ceased. He produced a long description together with an album of chromolithographs that capture evocatively the strange, almost unreal landscapes that met his eyes. The Royal Society in London appointed a committee that produced a second account. Its members did not visit Krakatoa but collected a wealth of data which they sifted and published in a report of great clarity. Apart from the scientific publications, there are numerous accounts, mostly in Dutch, from survivors together with the logs of several ships that were passing through the Sunda Straits at the time. The most remarkable of these is that of Captain W. J. Watson of the British ship *Charles Bal*. Captain Watson and his crew were the nearest witnesses of the eruption. Many other observers recorded the times of particular manifestations of the event and these can be related to a remarkable graph produced by the pressure gauge at the Batavia (now Djakarta) gas works during the second day (27 August) of the eruption. This gauge also recorded the times of numerous explosions and enable a very accurate chronology of events to be built up, as is detailed in a book to be published by the Smithsonian Institution in Washington, which provides the basis of the general chronological diagram in Figure 34.2.

Low in the sequence of volcanic deposits on all three main islands of the Krakatoa group (see Figure 34.1), and lying beneath the 1883 deposits, there are andesitic lava flows, and layers of pumice and scoria. These are the remnants of a large volcano that had existed in prehistoric times and probably reached a height of some 2000 m. After a major eruption, probably not unlike that of 1883, the volcano collapsed leaving only the three main islands together with a small islet. A new cone, Rakata, then formed on the southernmost island, built up of lava flows and fragmental volcanic debris and it eventually reached a height of some 800 m. In contrast to the prehistoric cone, Rakata was built principally of olivine basalt. Two smaller cones of andesite then developed to the north of Rakata. These were Danan, which reached 450 m, and Perboewatan, which was only 120 m. The three cones later coalesced to form a single island measuring 9 x 5 km. An andesitic

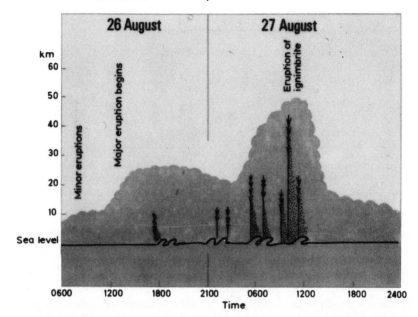

Figure 34.2 *Chronology of eruptions – 26 and 27 August, 1883. Major eruptions, ash fall and tsunami are shown diagrammatically (tsunami not to scale) (Based on Self and Rampino, 1981)*

flow is known to have erupted from Perboewatan in 1680 but for the next 200 years no volcanic activity of any kind was observed.

This, then, was the situation in 1883. The islands were covered with a luxuriant forest and occupied by fishermen from time to time. The local population had no reason to rear the peaceful island that lay athwart the Sunda Straits ... a landmark well known to the numerous ocean-going ships that were funnelled through the straits *en route* for the ports of South East Asia and the Far East.

On 20 May, 1883, Krakatoa once again became active. The initial eruptions were not particularly violent although towns on the nearby coasts of Java and Sumatra were showered with ash, earthquakes were felt from time to time, and the larger explosions were heard in Batavia, 160 km to the east. After a week, a party from Batavia landed on the island, and saw that the volcanic activity was confined to the northern-most cone of Perboewatan, and that the ash column had risen to 3 km. Activity varied in

Figure 34.3   *Verbeek's depiction of the view of Rakata from the north, after the 1883 eruption. The cliff is about 4 km long with a maximum height of 800 m. (Credit: Natural History Museum)*

intensity through June, July and August, and on 11 August another party landed on the island – the last to visit before the major eruption – and noted three active vents including Danan, and that there were 11 other visible steam vents. All vegetation had been destroyed and the islands were mantled with half a metre of fine volcanic ash.

On the morning of 26 August all was relatively quiet. Then just after noon, a series of huge explosions heralded the final catastrophic phase that was to last about two days. At about 1400 hours an intense black cloud rose to a height of some 27 km. Numerous sharp explosions gradually increasing in intensity, followed until at about 1700 hours, it seems that part of the volcanic cone collapsed. The first tsunami were then set in train, killing many people at Merak (Figure 34.1, inset) on the Java shore. An intense blackness descended over the straits and the adjacent coasts of Java and Sumatra and was to remain for the best part of two days.

Explosions continued all night, the incessant noise and rattling of doors and windows keeping everyone awake in western Java. Some explosions were so great that they snuffed out gas lamps in Batavia, while the larger ones were heard in Singapore, 1400 km away. The first of the four culminating explosions, which almost certainly marked the final collapse of the volcano and generated the largest of the tsunami, occurred at 0530 hours in the morning of 27 August. The second was just before 0700 hours and the third and largest was at about 1000 hours. This third detonation, probably the loudest and most powerful natural explosion in recorded history, was heard more than 4800 km away at Rodrigues and generated an atmospheric pressure wave that was recorded on barograms all over the world. Indeed, many barograms recorded the wave seven times as it passed back and forth three times to its origin, from its node point over Central America. Volcanic ejecta, driven to heights of around 50 km, covered an area of some 750 000 sq. km. About half an hour after the third of the explosions a tsunami which in places reached a height in excess of 40 m, hit the Javanese and Sumatran coasts. More than 300 towns and villages were totally or partially destroyed and the wave penetrated up to 11 km inland. The final death toll was more than 36 000 – the majority drowned, but many, particularly in southern Sumatra, were overcome by the fall of hot ash and the accompanying noxious gases.

The fourth great explosion occurred just before 1100 hours although strangely this time there was no attendant tsunami. Explosions continued for the rest of the day and the following night, but with diminishing intensity, and by 28 August it was all over. There was a little activity from time to time later in 1883 but it was relatively minor. The volcano was spent. Although activity had started in May, almost all its energy was dissipated in two days and the greater part on one day, 27 August.

Only about one third of Krakatoa remained after the eruption (Figure 34.1). The main peak of Rakata was truncated (Figure 34.3), and where the northern part and the cones of Perboewatan and Danan had once stood there was now more than 250 m of sea. Sertung and Rakata Kecil Islands were somewhat enlarged by thick new deposits.

About 20 cu. km of material had been ejected from Krakatoa. The bulk had fallen within 20 km, but a small proportion had been dispersed far more widely, some, indeed, being distributed over much of the Earth. About 95 per cent of the ejecta consists of pumice, a glassy rock froth produced by the rapid vesiculation then freezing of a molten rock. Some pieces of glassy obsidian occur in the pumice deposits together with about 5 per cent of rock fragments which are demonstrably part of the original cones. This observation is of importance when one considers the mechanism of the eruption. Many early geologists, including J. W. Judd, who wrote the geological section of the Royal Society report referred to earlier, considered that the northern part of the island had been simply blown away in an explosive decapitation of the

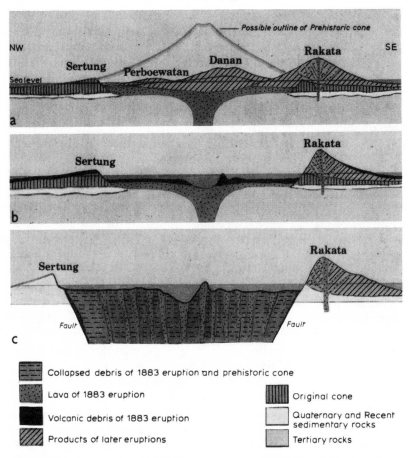

Figure 34.4 *Sections from NW to SW across Krakatoa* (a) *Situation before 1883 eruption (after Verbeek).* (b) *Situation after 1883 eruption (after Verbeek). Note implied paucity of 1883 debris.* (c) *Situation after 1883 eruption interpreted in terms of the subsidence of the cauldron. Note abundance of collapsed debris or original cones (after Holmes, 1965)*

island (Figure 34.4a and b), and that part or all of this explosive activity had been caused by sea water gaining access to the inner parts of the volcano. Had this been the case, a much greater proportion of the volcanic products of 1883 would have consisted of remnants of the original cones, rather than pristine glassy pumice which is clearly the product of fresh, primary magma.

It is now generally accepted that the northern part of Rakata disappeared because of the collapse of the volcano and the formation of a caldera, in the manner described by Howel Williams in his classic paper "Calderas and their collapse" published in 1941. Calderas form when the superstructure of a volcanic edifice collapses along a circular fault or group of faults, as the magma reservoir beneath the volcano becomes depleted either because the magma is withdrawn or is ejected at the surface. Such calderas are to be seen at many volcanic centres today and such a mechanism implies that the bulk of the cones of Perboewatan and Danan, and the northern half of Rakata, now lie beneath the sea below their former sites (Figure 34.4c), thus accounting for the paucity of old volcanic material among the products of the eruption of 1883.

Further insight into the nature and mechanisms of the eruption may be gained from a study of the eruptive products and their stratigraphy. Figure 34.5, based on Figure 2 of S. Self and M. R. Rampino (*Nature*, vol. 294, p. 699), gives the succession of the 1883 ejecta which is about 70–80 m thick over the islands. The lower part, representing the materials ejected on the first day, 26 August, is of airfall tuffs comprising numerous layers of ash, each approximately representing the products of one explosion. The material is chiefly pumice, and sometimes larger fragments are concentrated at the bottom of a layer having been winnowed as they fell through the air.

Some of the deposits are welded – the erupted fragments were so hot that they fused after accumulation. This indicates not only high temperatures, but also particularly rapid rates of accumulation. Most of the lower airfall deposits are rather coarse grained, suggesting that they are not the products of "phreatomagmatic eruptions" – that is to say, produced by the interaction of the magma with sea water – for such deposits are usually characteristically fine grained. The welding would also seem to preclude phreatomagmatic activity.

The top part of the succession, the products formed after 0500 hours on 27 August, 1883, are rather different, and are composed of four major units, in all 50–60 m thick. The materials of each unit, again principally pumice, show no signs of size sorting. The presence of these units leads to an interpretation and understanding of the eruption that was not available to Judd, Verbeek, or

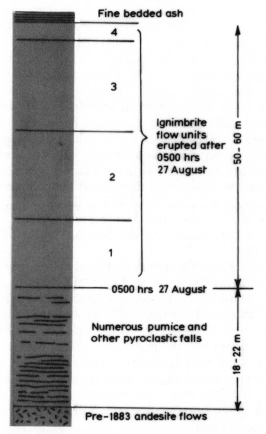

Figure 34.5 *Diagrammatic section through the 1883 pyroclastic deposits (Based on Figure 2 of Self and Rampino, 1981)*

their contemporaries, for these deposits are "ignimbrites" (see below), whose nature and mechanism of formation were not understood until many years later.

In some industrial systems, the process of fluidisation is used to transport solid material. Air is forced into finely divided solids such as sand or coal so that the individual particles are buoyed up and will flow like a liquid. Some types of volcanic eruption behave as natural fluidised systems and are responsible for the deposition of many of the most voluminous volcanic products. Eruptions of this type were first witnessed in 1902 emanating from the volcano Pelée in Martinique in the form of *nuées ardentes*. These incandescent volcanic clouds are fluidised systems in which the buoyancy is attained by the gases initially incorporated into the flow, by expanding air engulfed in the flow as it moves, and by gases released from the solid particles within the flow. Such flows ("pyroclastic flows") can not only carry the finer solid material but also huge blocks. They start within the volcano by the rapid vesiculation of magma, that is the rapid foaming of the magma caused by release of dissolved volatile components when the confining pressure is released; the usual analogy is to the frothing of a bottle of Champagne or fizzy lemonade when the stopper is removed. The rapid increase in volume ejects the mass from the volcano, and the solid particles continue to release gas which effectively helps to keep the system mobile.

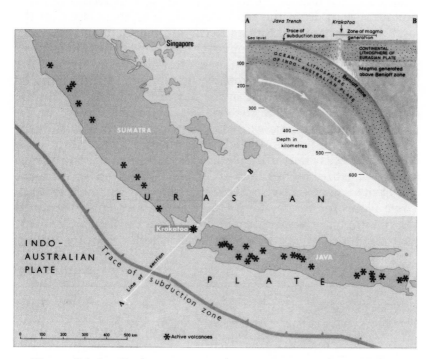

Figure 34.6   *Krakatoa in its plate tectonic setting. Vertical and horizontal scales of section are the same, but the sea-floor profile and the precise sites of magma generation have been exaggerated for clarity. The "Benioff zone" is defined by the loci of earthquake foci and marks the upper surface of the descending lithosphere plate*

The material may devolve directly from the vent as a dense, enveloping blanket of incandescent rock fragments and gas which either can be directed at great velocity laterally or can be driven upwards as a vertical eruption column. The column will eventually be overtaken by gravity so that it will collapse and fall back on the summit and flanks of the volcano, flow down them and spread out over the lower ground. The flow will continue to act as a fluidised system as long as it can continue to expand and to buoy up its constituent solid phase. When this is no longer possible, its load of blocks, ash and dust will be deposited as an unsorted mass.

The upper part of the 1883 Krakatoa deposits are of this chaotic type – known as ignimbrites – and the four units were probably deposited by four flows of pyroclastic material almost certainly erupted at the time of the four major explosions recorded on 27 August. Apart from the thick deposits now to be seen on the islands of Sertung and Rakata Kecil, and the south and east sides of Rakata, the flows spread out over the sea, in places filling it and forming temporary islands. These are shown on Figure 34.1 together with the overall probable distribution of the ignimbrite deposits. The islands to the north and east of Rakata Kecil were soon removed by marine erosion.

Because pumice is a rock foam of low density it will float. Ships in the Sunda Straits immediately after the eruptions, and for some days and weeks afterwards, reported large rafts of floating pumice more than three metres thick in places. These were so extensive along the Sumatran coast that rescue ships were unable to reach some places for several days. Ocean currents gradually dispersed the pumice but it was reported from all over the Indian Ocean for many months, before it was washed ashore or became water-

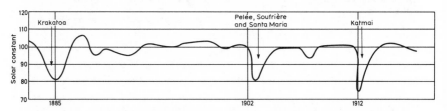

Figure 34.7 *Variation of solar constant with time showing the effect of the volcanic eruptions of Krakatoa; Pelée, Soufrière and Santa Maria in 1902; and Katmai in 1912 (after Kimball,* Monthly Weather Review *(1918), vol. 46, p. 355)*

logged and sank. A distinctive horizon of Krakatoa pumice from the 1883 eruption can now be recognised along many coasts around the Indian Ocean, and some splendid pieces from the coast of Kenya were recently acquired for the British Museum (Natural History).

So the disappearance of the northern part of Rakata is now ascribed to collapse during caldera formation. This collapse was certainly heralded by the four major explosions on the second day, concomitant with the eruption of the pyroclastic flows which deposited the ignimbrites, and it was the rapid loss of this material from beneath the volcanic edifice that must have led to its foundering in four stages. Three of the four major explosions were also accompanied by the largest of the tsunami, though one explosion seems not to have produced such a wave.

Exactly how the tsunami were generated is still far from clear. Submarine explosions, collapse associated with the formation of a caldera, and large volumes of ejected material falling into the sea probably contributed.

Two questions remain. First, why is Krakatoa where it is, and secondly, will it erupt again? The first question is readily answered in terms of plate tectonic theory. Running just to the south of the southern coasts of Java and Sumatra is a boundary between the Indo-Australian Plate, to the south, and the Eurasian Plate to the north (Figure 34.6). The Indo-Australian Plate is moving slowly northwards and being deflected downwards beneath the Eurasian Plate. The line of junction of two plates where one plate passes beneath another is known as a subduction zone, and most of the world's most active volcanic belts lie close to subduction zones. As the descending plate moves down the Earth's mantle it is heated and partially melted at depths of between 100 and 300 km. Because the newly melted plate material is less dense than the surrounding mantle it rises towards the surface where it may be erupted as lava. This process takes place along the margin of the plate, which accounts for the line of volcanoes above the descending plate. Throughout Java and Sumatra there are about 100 active volcanoes. Krakatoa being one of them. Because Krakatoa forms a small island it has the potential to generate tsunami, which is denied to most of the other Indonesian volcanoes which are well inland, and it is this situation more than any other that has undoubtedly given this volcano its special place in human history.

In 1927 Krakatoa once again came to life and a new island

Figure 34.8   *The boiler is all that remains of the steamship Berouw which was carried 2½ km inland by tsunami (Credit: Krafft)*

Anak Krakatoa (Figure 34.1), was built in a position once occupied by the northern part of Rakata. This island is now well over a kilometre across and in recent years has been very active. The materials being erupted at present are basaltic in composition and so have relatively low viscosities. However, the magma beneath the volcano will probably evolve chemically to compositions having higher viscosities, and it is then – to attempt to answer the second question – that dangerously explosive activity can be expected. At present it seems that this is unlikely to be in our lifetime.

One cannot leave the subject of Krakatoa without mention of the meteorological effects. On 27 August, 1883, the largest explosions drove fine volcanic dust high into the stratosphere, where it remained for several years. It was swept at first as a narrow band around the equator, but over several months the dust layer widened until it extended into moderately high latitudes including most of Europe and the United States. The optical effects caused by the dust were dramatic, including the production of extraordinary sunsets, haloes around the Sun (Bishop's rings), and the Sun and Moon appearing blue or green from time to time. In London a W. Ascroft of Chelsea made crayon drawings of the finest effects over a period of four or five years, eventually accumulating over 500 drawings. These are now deposited with the Science Museum in London. The Royal Society Krakatoa Committee used his more dramatic drawings as the frontispiece to its report.

A more profound effect of the presence of the dust in the stratosphere was on the weather. There was a marked decrease in the solar constant – the amount of solar radiation reaching the surface of the Earth (Figure 34.7) – for several years, resulting in much cooler conditions than normally prevail throughout the northern hemisphere. For the 1883 Krakatoa eruption at least, the correlation of indifferent weather with volcanic activity would seem to be proved beyond reasonable doubt. However, whenever there is a large eruption nowadays, such as at Mount St Helens in 1980, exceptional sunsets and inferior weather are predicted, but indubitable correlation can be established only rarely. The effects on the solar constant of the 1902 eruption of Pelée, Soufrière and Santa Maria in 1902, and the eruption of Katmai in Alaska in 1912, can be seen in Figure 34.7. For there to be any likelihood of

Figure 34.9   *View of the smouldering Anak crater (Credit: Krafft)*

weather patterns being affected volcanic materials must be projected into the stratosphere, so that only eruptions of an extremely explosive nature, and then only those resulting in the formation of a large eruption column, are likely to qualify. Basaltic volcanism of the Etna or Hawaii type, involving the gentle outpouring of relatively low viscosity basaltic magma, however voluminous, is not able to do this. Sufficient energy was certainly released in the initial explosive eruption at Mount St Helens on 18 May, 1980, to have driven ash into the stratosphere, but much of that blast was directed sideways as the northern side of the volcano collapsed, so probably muting the effect. Only very occasionally, it seems, are volcanic eruptions powerful enough to disrupt weather patterns to such an extent that the change can be distinguished unequivocally from the normal very variable and unpredictable patterns.

25 August, 1983

35

The Earth flexes its muscles

JAMES JACKSON AND ROBERT MUIR WOOD

A science of mountains is emerging, and is full of surprises. It shows, for example, that the Earth's continental crust stretches and contracts like a muscle.

Before geologists can hope to understand violent complex mountain ranges such as the Karakorum – the focus of attention for the Royal Geographical Society's International Karakorum Project 1980 – the beginnings of a science of mountains must be built around simpler ranges. One could argue that romantic poets had, up to the past decade, contributed as much to our understanding of mountains as geologists: inspiration, a vocabulary, and the knowledge that within their seeming randomness might lie buried structures. Modern theories of continental drift, in the form of plate tectonics, have helped explain such features as the Alpine–Himalayan mountain chain in terms of continental collisions but they have not advanced an understanding of the finer detail. For mountains are not products of random destruction, but are large-scale highly structured features, which have been awaiting some simple model to show not only how but also why they form. Such a model is now in the process of construction: to understand mountains it is necessary to return to the material out of which they were built and to study them soon after they begin to be born.

Plate tectonics emerged from the study of the oceans and showed the ocean-floor lithosphere – the Earth's upper layer – to be the top of a deeper level convection process. The creation and destruction of this lithosphere are processes of extraordinary simplicity. Except where two plates interact the ocean-floor is rigid and capable of moving for thousands of kilometres without suffering distortion. This rigidity is the most fundamental concept within plate tectonics.

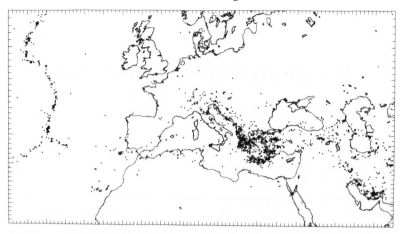

Figure 35.1   *Earthquake epicentres define plate boundaries in the Mid-Atlantic, but occur in a much broader region in the Eastern Mediterranean where continents are colliding*

   The relative displacements of the various plates are too slow to be measured directly and the boundaries between the plates are identified by the concentration of earthquake epicentres, resultant on the sudden movement of faults. At plate boundaries in oceanic lithosphere these earthquakes lie on narrow and distinct zones. Where two continents collide, however, this simplicity breaks down and the earthquake epicentres extend over regions hundreds of kilometres wide, and do not define a distinct plate boundary (Figure 35.1). The separation of aseismic (earthquake-free) rigid areas from narrow seismically active zones is in this case no longer obvious; can any simple pattern be retrieved?
   The strong differences between continental and oceanic crust are likely to have a determining influence on crustal deformation. Some continental crust dates back 3500 million years whereas the oldest oceanic crust was formed only 200 million years ago. The structure of the ocean floor is simple and consists of undeformed sediments lying above a thick pile of basaltic igneous rocks (crystallised from molten lavas) extruded at a mid-ocean ridge between plates. By contrast the continents contain rocks that have been repeatedly subjected to great changes in temperature and pressure, and have been involved in a complex history of deformation leaving them riddled with lines of weakness. Continental crust should be weaker than oceanic crust as it contains

abundant granitic and silica-rich rocks that deform at lower temperatures than the iron- and magnesium-rich basalts of the ocean-floor. Furthermore continental crust is lighter than the rocks in the upper mantle beneath it; this buoyancy tends to prevent one continent being pushed under another. In contrast old oceanic crust is denser than the underlying mantle and sinks at convergent plate boundaries causing earthquakes as deep as 700 km. In continental collision zones earthquakes deeper than 40 km are rare, and the principal reaction of continental crust to horizontal compression is to thicken, pushing up the Earth's major fold mountains.

Figure 35.2   *Part of the Zagros range in Iran, north-east of Shiraz — fold mountains that result from the collision of Arabia with Iran*

Fold mountains are the result of plate interactions, but the way
in which they form is complicated and not well understood. The
Alpine–Himalayan ranges are an important band of crustal
shortening (thickening) many hundreds of kilometres wide; they
are ultimately caused by the collision of Africa, Arabia and India
with Eurasia. Is plate tectonics helpful for describing and predicting the deformation of so vast region? The observed faulting
could be considered as the result of relative movement between a
large number of small rigid blocks, but these become so numerous
and so arbitrarily defined in the presence of the region's diffuse
seismicity that a description in terms of "mini-plates" becomes too
complicated to be useful.

Satellite photographs reveal that Asia is scarred by a number of
giant linear strike-slip faults (Figure 35.3) similar to the San

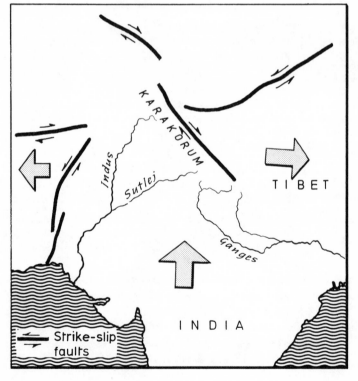

Figure 35.3   *Blocks of continent slide past each other in
Asia, moving horizontally (that is, along the surface of the
Earth) on giant linear strike-slip faults*

Andreas Fault in California. The predominantly horizontal motion on these faults has caused blocks of continent to slide past each other, in some cases by more than 100 km. Sometimes earthquakes concentrate on these faults and "fence off" areas of low seismicity. However, where the Earth's crust is shortened or extended (stretched) faults are activated over a wide area. Big strike-slip faults appear to absorb some of the crustal shortening by sliding large regions of continental material sideways out of the collision zone thereby avoiding excessive thickening.

It is difficult to determine the rates of movement involved in continental collision, though in some parts of the world it is possible to "read" collision speeds from the magnetic "stripes" – bands of alternating magnetisation – found in the ocean-floor trailing a continent. From the Indian Ocean floor we know that India is now moving into Asia at some 30 mm/year. The remanent magnetism of rocks provides a means for establishing the latitude of continental fragments in the past, and is useful for testing models of the history of continents. Such work demonstrates that Eurasia has shortened by 1500–2500 km since the collision with India began about 40 million years ago, and confirms that fragments of Gondwanaland (the supercontinent of South America, Africa, Antarctica and Australia that started to break up about 200 million years ago) exist in central Iran and Afghanistan.

To get more detailed information it is necessary to study those earthquakes that represent the present-day faulting processes in young mountain belts. If the yongest and simplest zones of continental collision can be understood then the greater confusions of the older mountain belts may themselves be unravelled. Large earthquakes provide the best information as most seismic slip happens during the largest shocks, while the more frequent smaller earthquakes may help to indicate previously unrecognised faults.

Fault movements associated with the major earthquakes are studied using the permanent global network of seismograph stations. The radiation pattern of seismic waves from large earthquakes may be analysed to deduce the type of orientation of faulting involved (see Box). Sometimes a fault rupture appears at the surface, confirming this deduction. Such ruptures are often a small part of a much larger fault structure visible on satellite photographs. By monitoring the aftershocks following a major earthquake it is possible to find the approximate area of the fault plane that moved. This coupled with the size of the main shock can indicate the amount of movement. If the earthquake history of the fault is known the overall rate of movement can be estimated. Some continental fault systems move several centimetres a year — rates comparable with those of plate displacement.

The problem . . .

The focal depths of continental earthquakes provide particularly useful information if they can be reliably estimated, although routine determinations of focal depth can be in error by 100 km for small shallow shocks in the Alpine–Himalayan belt. This misidentification of shallow earthquakes as deep subcrustal shocks has in the past led geologists to devise inappropriate models of continental collision zones, based on oceanic tectonics, in which material is pushed back into the mantle. Deep subcrustal earthquakes are in fact rare in true continental collision zones, though some have occurred beneath southern Tibet.

Long ago geologists realised that many of the classical features of mountain belts, such as folds and near-horizontal thrusts faults (on which one unit of crust has been pushed on top of another) imply considerable horizontal shortening. As these features are visible at the surface usually in very thick sediments deposited in shallow water, it has long remained a mystery as to what happened to the harder crystalline continental "basement" on which these sediments were once deposited. Very often the sediments involved are separated from their basement by a layer of salt or shale that becomes ductile under pressure and acts as a lubricant, so that structures above need bear no relation to those below. Exploration deep below the surface in the Rockies and Appalachians has confirmed this decoupling; there folds and thrusts lie above an apparently undeformed basement (Figure 35.4). What then happens to this basement if its sedimentary cover is substantially shortened? As its buoyancy should prevent it from being pushed down into the mantle, crustal thickening might take place instead. Yet observations in the Alpine–Himalayan belt show that continental blocks often move sideways on strike-slip faults to avoid moving down against gravity as would be necessary to over-thicken continental crust.

The problem of too much sediment and too little basement is the central "space problem" of mountain-building. A partial solution has come, ironically, not from the study of mountains but from the

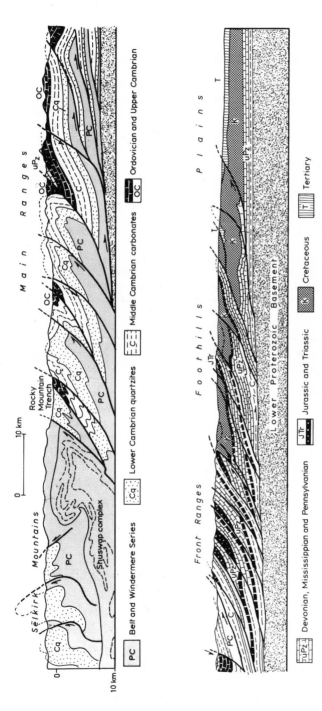

Figure 35.4 A section through the Rocky Mountains (a little north of the Canadian Pacific railroad) reveals a series of curved faults perpendicular to the main axis of the mountains. The crumpled sediments are decoupled from the underlying basement, which has remained undeformed (after Price and Mountjoy)

Faulting and seismic waves

The seismic waves from a large earthquake, with magnitude greater than about 6 on the Richter scale, will be recorded by seismographs over most of the Earth's surface. The shape of the waves contains information from which the type and depth of faulting involved in the earthquake can be deduced. Below are idealised seismograms for a station in Ireland recording an earthquake in the Zagros mountains of Iran. The first three waveforms are for the three main types of fault motion. Note how the polarity of the onset of ground motion is sometimes up and sometimes down. This depends on both the direction of the station relative to the orientation of the fault and the type of faulting involved. By examining polarities at many stations widely distributed round the globe the orientation and the sense of movement of the fault can be found. The second set of three seismograms shows the change in shape of the waveform as the focal depth of the earthquake changes. The polarity of the onset is the same for all three as this depends on the type and orientation of faulting, which is the same in all cases. However, the pulse shape clearly changes and can be used to estimate the depth at which faulting occurred.

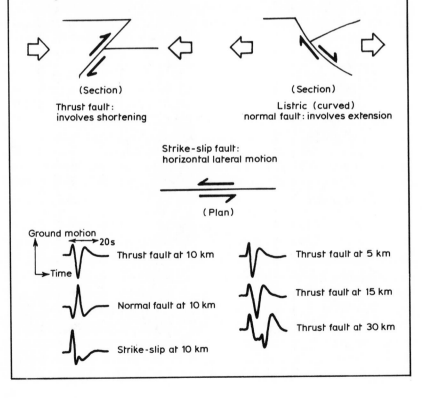

(Section)

Thrust fault:
involves shortening

(Section)

Listric (curved)
normal fault: involves extension

Strike-slip fault:
horizontal lateral motion

(Plan)

Ground motion

20 s

Time

Thrust fault at 10 km

Normal fault at 10 km

Strike-slip at 10 km

Thrust fault at 5 km

Thrust fault at 15 km

Thrust fault at 30 km

study of the hollows of the Earth's surface (*Nature*, vol. 282, p. 343). In 1976, American geophysicist J. Helwig pointed out how the "space problem" in shortened mountain belts is lessened if the basement underneath the folds is thin to begin with; subsequent shortening simply returns it to a normal thickness. In 1978 Dan McKenzie, of Cambridge University, showed that the subsidence of continental sedimentary basins can be quantitatively described by a simple model in which the lithosphere is rapidly stretched and thinned, sinks quickly as its base is replaced by hot dense mantle material from below, and then subsides more slowly as the whole lithosphere cools in a manner similar to that of a mid-ocean ridge. The initial stretching of the basement at shallow depths is thought to occur by motion on curved (listric) faults. Thereafter stretching ceases and subsidence continues without further faulting.

This model originated from a study of crustal thickness and heat flow below the Aegean Sea, where crust is still being stretched by normal faulting (that is, with vertical compression) and is still sinking. The model also appears to account well for the history of subsidence of the North Sea and many of the Carpathian basins in Eastern Europe, all of which show an initial episode of normal faulting followed by subsidence with no further faulting. The gross features of the same model will account for the stretching and subsidence preceding the break-up of a continent to form an inactive margin such as the east coast of North America; in this case the stretching has gone too far, and an ocean has opened up down the middle of the basin. All thick sediments near continental margins probably form on thin extended basement which was stretched by normal faulting. Such thinned basement exists in both the North Sea and the Vienna Basin. The Aegean Sea, however, is in an earlier stage of development, with high heat flow, and active normal faulting.

... and a solution

There is thus every reason to think that Helwig's suggestion has some validity, and that basement under the thick piles of sediment that make up fold mountains had been thinned by normal faulting before the onset of collision with another continental margin. The faulting was probably of a low-angle listric nature, spread over a region more than 100 km in width, as observed in the Great Basin of the western US, and in the continental margin off western France (*New Scientist*, 6 November, 1980, p. 362).

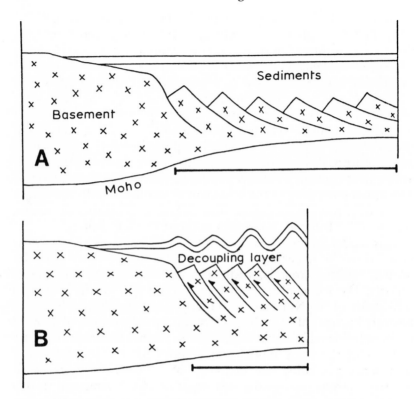

Figure 35.5   *In the process called reactivation,* old *normal faults (that is, those that involve extension of the continental crust) act as thrust faults (those that involve crustal shortening, or thickening). In A, sediments have been deposited on a basement which has been stretched and thinned by listric (curved) normal faulting. In B, the motion is reversed on these faults, and the sediments take up this shortening by folding*

During the early stages of continental collision the basement may take up this shortening by reversing the sense of motion on the pre-existing normal faults, and using them as thrust faults. In this way the basement tends to return to its original thickness while the overlying sedimentary column, having been deposited on a stretched basement, is forced to take up the shortening by folding. Early in the collision the compression can be achieved without thrusting continental basement into the mantle, or thickening it beyond its original (unstretched) state. This hypothesis provides at least a partial solution to the "space problem".

A likely region for testing the hypothesis is in the Zagros mountains of Iran – a young fold mountain belt currently shortening as a result of the collision between Arabia (and Iraq) with Iran. The structure of the Zagros is simple: a thick sequence of sediments deposited in shallow water on a continental shelf has warped into gentle folds during the past 2 million years. Because the mountains are still seismically active we can use the earthquakes to provide information on faulting deep below the surface. Long linear fold axes, and seismic activity showing thrust faulting of the same orientation, indicate there is to be a general NE–SW shortening. The relative motion between Arabia and Iran is perpendicular to the trend of the belt, and no blocks appear to be moving sideways from the collision zone.

The region is seismically very active, with earthquakes spread over an area about 300 km wide, and with a distinct north-eastern boundary which in turn marks the edge of the Zagros sedimentary trough. These earthquakes are probably all shallower than 40 km and do not appear to increase in depth towards the north-eastern margin of the belt. Consequently there is no evidence for seismic shortening on a single gently-dipping plane, as is the case at plate margins where oceanic crust is pushed back into the mantel beneath. Seismic radiation patterns from Zagros earthquakes consistently show thrust faulting with comparatively steep (40–50°) fault planes. This is in marked contrast to the less steep (0–10°) thrust faulting which occurs in areas where oceanic underthrusting of continents is, or was, taking place, such as in Alaska and the Eastern Himalayas. The seismic activity of the Zagros Fold Belt reveals steep faulting on a large number of faults distributed across the whole width of the belt.

The routine determinations of focal depth, which use the arrival times of seismic waves reaching the global network of seismograph stations, rather than any study of the waveforms themselves, are not sensitive enough to distinguish the small (5–10 km) changes in depth which may greatly alter the geological significance of an earthquake. In the Zagros it is particularly important to know whether the large shocks represent deformation in the basement or in the folded sedimentary column. At the base of the sedimentary column, above the crystalline basement, there are several thick salt deposits. Any faulting in the basement is unlikely to propagate to the surface through these ductile layers. This may explain the curious observation that no surface ruptures have ever been seen following a large earthquake in the Zagros, whereas similar size shocks in central Iran, where there is no salt at depth, commonly have surface ruptures several tens of kilometres long. The aftershock distribution of large Zagros earthquakes often spreads out over 30 km, and it is unlikely that faulting of this dimension could be contained within sediments 6 km thick without showing some surface faulting. The most convincing evidence that the faulting is actually within the basement comes from an examination of the seismic waveforms from Zagros earthquakes. In three areas of the Zagros this suggest that earthquakes occurred at depths of 12–15 km. In each case this is below the anticipated thickness of sediments. It seems likely, therefore, that it is the basement beneath the Zagros Fold Belt that is deforming on steep thrust faults.

During most of the past 200 million years, what is now the folded belt of the Zagros was apparently a slowly subsiding continental margin. If the present-day thrust faulting is happening on the old normal fault surfaces which originally caused the basement extension, it would explain the high (40–50°) rather than low (0–10°) angles of the fault dips. Most conclusively, the dips of the fault planes in the Zagros are similar to those in the Aegean Sea, which is currently extending by normal faulting.

Such reactivation of old faults is a common and important phenomenon, and the rejuvenation of old basement normal faults as thrusts has recently been reported from Chile. The Ramapo Fault in New York State, currently active as a thrust fault, has been active at several times in its past with different senses of motion. Studies of active tectonics in Asia show many examples in which old mountain belts and zones of weakness on continents have become reactivated during the collision of India with Asia, while still older areas remain undeformed. Reactivation of old faults is probably as fundamental to continental tectonics as rigidity is to the behaviour of oceanic plates.

Towards the Karakorum

An objection to this proposal that the continental crust stretches and contracts like a muscle, has come from geophysical exploration in old fold mountains, such as the eastern Rockies, where undeformed basement is found dipping gently beneath folds in front of the main mountain axis. Yet in these older mountain belts the amount of shortening involved is considerably greater (about 160 km, or 50 per cent, in the Canadian Rockies, and probably more than 200 km in the Appalachians) than that which has so far taken place in the juvenile Zagros (about 20–50 km or 20 per cent).

The contraction mechanism we have discussed will presumably work until the reverse motion on old normal faults restores the basement at least to its original thickness. Thereafter some other process must occur. It is probable that the sediments of the folded belt of the Zagros are still more or less above the original basement on which they were deposited. But as shortening continues, the folded cover will start to migrate south-west over the undeformed Arabian shield, separated from its basement by the decoupling basal salt layer, just as the folds of the Rockies and Appalachians are no longer above their original basement which was left behind in the core of the belt. How long this reactivation process can continue is not clear. It is unlikely to account for many of the features of older mountain belts which involved considerably greater shortening. However, even if operative in only the early stages of collision, it helps solve the first problem of mountains, the "space problem" encountered in trying to unravel fold and thrust belts. It also provides a beginning from which to try to understand the processes operating in more violent and tortured mountain belts such as the Karakorum.

11 December, 1980

36

Ice ages and continental drift

CHESTER BEATY

Massive glaciations on Earth can occur only if the arrangement of the continents is suitable.

Speculation about climatic change – particularly change supposedly leading to massive glaciation – has long been a preoccupation of Earth scientists. At the same time, sensational predictions of impending climatic doom have confused and, on occasion, frightened the general public. All too frequently in recent years, one "expert" or another has declared that we are on the brink of some sort of atmospheric disaster, that massive glaciation or global desiccation is imminent. With justification, at least some people are beginning to have doubts about much of the current meteorological wolfcrying, and there is reason to suspect that a new ice age is not, after all, just around the corner.

We can make sense out of the problem of glacial causes by constructing a relatively uncomplicated model incorporating the various natural factors that seem to interact so as to give rise to episodes of major glaciation. And the model to be formulated is one in which climatic change, as such, is not required. Rather, development of large-scale continental glaciation leads inevitably to changes in atmospheric characteristics, some of the evidence of which has probably been misinterpreted.

The basic environmental condition for initiation of glaciation is almost ludicrously simple: More snow must fall and accumulate during the winter than will be disposed of by ablation in the following summer. Persistence of this state of imbalance for a few years will, of necessity, lead to the beginning of a period of glaciation. What is lacking in such a simplistic account is, of course, an examination of the conjunction of multiple factors giving rise to sustained glaciation on a continental scale.

There are many potential causes of ice ages, including such standard items as climatic changes – that is, changes in the atmospheric circulation – and mountain-building, fluctuations in cosmic and/or galactic parameters and even possible variations in The Newtonian gravitational constant, G. However, while there may well be numerous other factors at play, a workable model of glacial causes can be put together utilising some or all of the following seven, not necessarily arranged in order of relative importance: (1) solar output and variability; (2) changes in the Earth's orbital geometry; (3) orogenic activity and continental uplift; (4) volcanic activity; (5) variations in the Earth's surface albedo; (6) fluctuations in oceanic characteristics; and (7) changes in continental locations (continental drift). The scheme that emerges seems satisfactorily to provide an explanation for most of the geological evidence of past glaciations (and *geological* evidence, it should be kept in mind, is all we have; the atmosphere, unfortunately, does not leave a direct record of its past performance).

The model I describe here is founded on the assumption that the primary geophysical requirement for continental glaciation is the presence of large land masses in sufficiently high latitudes to catch and hold a lot of snow. Given this basic geographical situation, interaction of a number of the other factors leads inexorably to glaciation, with glacial-interglacial alternations controlled by variations in the Earth's orbital geometry. Since North America and Eurasia are now in the required high latitudes, it seems that the glacial-interglacial cycle on this planet will prevail until these major land masses have been moved into lower latitudes. The Earth, in short, is currently locked into an ice-age condition.

How does the model work? Available evidence (including especially the palaeomagnetic record) strongly implies that major glaciations have occurred only when the continents have been in high latitudes. Accordingly, continental drift (*à la* contemporary plate tectonics–sea floor spreading paradigm) assumes a fundamental role, from time to time during Earth history moving major land masses into appropriate positions. A glacial episode is then believed to be triggered by an increase in the Earth's surface albedo (the efficiency with which incoming radiation is reflected) which, in turn, leads to a lowering of tropospheric temperatures and a concomitant increase in snow fall in certain critical areas. A self-sustaining feed-back mechanism is thus created. Increasing surface albedo favours further temperature decline, with still more

precipitation falling as snow. In the right circumstances, continuation of this pattern for a few decades or hundreds of years propels
the Earth into an ice age.

The requisite albedo increase (the initial "trigger") is here held
to be brought about by the chance coincidence in time of the
effects of two other causal factors. First an increase in explosive
volcanic activity can produce an intensified volcanic "dust veil"
and resultant cooling. Secondly, a slight decrease in the total solar
radiational output associated with depressed sunspot activity, can
lead to additional cooling. The surface-albedo "trigger" is thus
activated by both internal and external stimuli.

Is there reliable evidence that such a postulated sequence of
events can indeed happen? The so-called Little Ice Age of the
17th–19th centuries provides a probable historical example.
Sunspot activity was drastically reduced from 1645 to 1715 (the
Maunder minimum), explosive vulcanism was widespread and
frequent, and there was a noticeable (and documented) increase in
the regional snow cover and lowering of the snowline in parts of
north-eastern Canada (and probably elsewhere as well). Temperatures dropped on a world-wide scale, and the stage seemed set, so
to speak, for initiation of another, or the next, ice age. Yet
atmospheric temperatures generally rose after about 1730 and
have tended to fluctuate above and below long-time mean values
since then.

A combination of factors

Evidently the "right circumstances" must include operation of at
least one of the other causal factors. The changing orbital
geometry of the Earth is a likely candidate. As was demonstrated
some time ago by C. Troll and M. Milankovitch (and, more
recently, by many others), small but systematic variations in the
inclination and orientation of the Earth's rotational axis, coupled
with changes in the eccentricity of its orbit, lead to fluctuations in
seasonal insolation (solar heating) received at any given latitude
on a cyclic, recurring basis. Such changes, predicated on accepted
principles of celestial mechanics, must occur independently of any
absolute variation in total solar radiational intensity.

The geometry of the Earth–Sun system is such that the Earth is
at present receiving relatively high wintertime insolation in much
of the Northern Hemisphere; 8000–18 000 years ago, wintertime

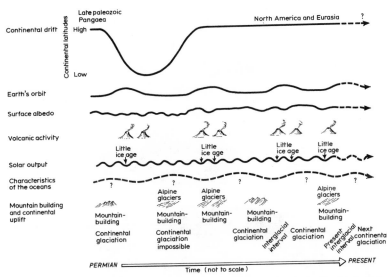

Figure 36.1 *The pattern of changing geophysical and astronomical features over the past millennia, related to the changing pattern of glacial/interglacial climate*

insolation in the north was significantly lower. The warming effect of greater wintertime insolation is apparently sufficient to overcome the comparatively short-time cooling induced by an increase in volcanic dust and a slight decrease in solar output. "Little" ice ages may develop, as happened a couple of hundred years ago, but big ones can't.

A return to conditions of widespread continental glaciation must await the next period of low wintertime insolation in the Northern Hemisphere. This will occur in about 8000–10 000 years, and since that era will also be a time of low summertime insolation, the glaciation that should then ensue will probably be much more extensive than the most recent advance which ended only 10 000–12 000 years ago.

I have omitted from consideration in this model of glacial causes the potential role of orogenic activity/continental uplift, and the possible effects of changes in oceanic characteristics. The omission is deliberate. While at the right time and place mountain building may lead to intensification of pre-existent glaciation (and certainly could generate local alpine glaciers), there is little geological evidence suggesting a cause-and-effect relationship between past

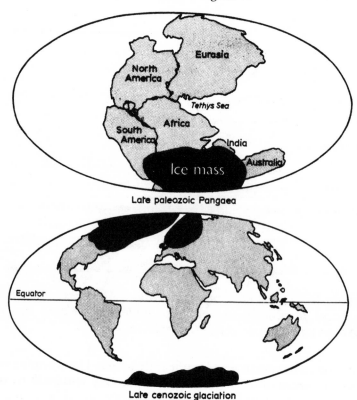

Figure 36.2   *Full ice ages can only occur if there is land near the poles*

episodes of orogeny and the onset of major ice ages. In the same vein, changes in oceanic characteristics, as revealed by the burgeoning deep-sea core record, are here regarded as more indicative of the *results* of glaciation than of its causes. On the glacial–interglacial time scale, the atmosphere seems to have led and the oceans to have followed. Significant and long-lasting changes in oceanic circulation patterns, brought about by continental drift, simply can not occur rapidly enough to have figured prominently as *initiators* of glaciation. A re-assessment may be necessary in the future, but it is concluded at present that these two latter factors occupy only secondary, not primary, positions in the glacial equation.

A fundamental aspect of this model is the proposition that a climatic *potential* for glaciation has always existed in the higher latitudes. Since climatic belts on this planet are arranged in a latitudinal zonation (and it is difficult to imagine a situation in which gross climatic distribution could be otherwise), major glaciations can develop *only* when large land masses are at high latitudes. This basic tenet of geologic history obviously was appreciated long ago by Alfred Wegener and, it should be noted, by only a very few others. One of Wegener's great advantages over most of his contemporaries lay in his realisation that on a round Earth heated by the Sun it is easier to move continents than it is to shift climatic belts.

Recent interpretations of the environmental implications of past and present distributions of certain climatically sensitive sediments, by G. E. Drewry, T. S. Ramsay, and A. G. Smith (*Journal of Geology*, vol. 82, p. 531) and W. A. Gordon (*Journal of Geology*, vol. 83, p. 671) have persuasively indicated that patterns of atmospheric circulation, and therefore the locations of the general latitudinal climatic zones, have remained more or less the same throughout most of the past 500 million years. The present model of glacial causes does not demand unreasonable behaviour on the part of the atmosphere and meshes well with the global pattern projected by these studies.

7 December, 1978

37

Impact craters shape planet surfaces

RICHARD GRIÈVE

Evidence is growing that the collision of planetary material with the Earth can profoundly affect local geology, and that impacts of very large meteorites may have influenced the evolution of the Earth and the life that exists upon it.

Craters formed by the impact of extraterrestrial material are not only a common phenomenon throughout the Solar System but they also once rivalled volcanism as the most important geologic process shaping the earliest surfaces of planets. The spacecraft images returned by the missions to the Moon, Mars, Mercury and, most recently, the icy satellites of the giant planets Jupiter and Saturn, have revealed extraordinary details of planetary surfaces. In many instances, the surfaces have been literally covered by impact craters ranging up to and over a thousand kilometres across. The familiar facial features of the "Man in the Moon" are, in fact, large impact basins filled by dark, basaltic lavas. These basins and by far the majority of the impact craters date back some 4000 million years and were formed by the bombardment of debris left over from the formation of the planets.

Planetary bodies such as the Moon, Mars and Mercury have thick, stable outer shells, or lithospheres, and have preserved portions of their early cratered crust. By contrast, the Earth's lithosphere is thin and is divided into a system of interlocking plates that are constantly in motion. The tectonic movement of these plates, and the action of erosion, results in a surface that is constantly being reworked and changed. Thus the Earth has retained no record of its earliest history of impact cratering.

The Solar System is now a considerably quieter place than during the first 600 million years of violent bombardment. Nevertheless, interplanetary material continues to bombard Earth, but at a reduced rate and in smaller quantities. About 100 000 kg (100 tonnes) of extraterrestrial material now enters the Earth's atmosphere each day. Most of this material is cosmic dust which burns up during its passage through the Earth's atmosphere.

On geologic time scales of millions of years, however, larger objects collide with the Earth. These are the Apollo objects, which are a family of interplanetary bodies whose orbits cross that of the Earth. Current estimates place the number of Apollos with diameters greater than a kilometre at about 1200. Although the chances of an Apollo body and the Earth being in the same place at the same time are small, it has happened in the past and undoubtedly will in the future. The evidence of past collisions is seen as terrestrial impact craters.

To date, approximately 100 craters larger than 1 km in diameter have been recognised on Earth (Figure 37.1).The majority have been found on the geologically old and central stable regions of the continents, such as the Canadian Shield and the associated flat-lying, tectonically undisturbed platform sediments of the Interior Plains. The relatively small number of known craters reflects not only the ability of the highly active Earth to remove the scars of impact, but also just how difficult it is to recognise eroded, degraded and sometimes buried impact craters. From a statistical analysis of the number and ages of the known

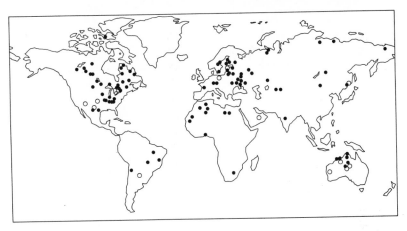

Figure 37.1   *Known terrestrial impact craters. Open circles are small (less than 1 km), very recent craters with meteorite fragments and shock features. Black dots are larger, older structures with shock features*

craters, it is estimated that a 10- to 20-km crater is formed somewhere on the land surface of the Earth every few million years.

Large-scale impacts are extremely rapid, dissipating huge amounts of energy. The average speed with which Apollo bodies hit the Earth is about 25 km/s. Speeds of this magnitude have little meaning in everyday life. Consider, however, that if you were on a flight from Ottawa to Montreal (160 km) and an Apollo passed overhead just as the plane was beginning its take off, then the

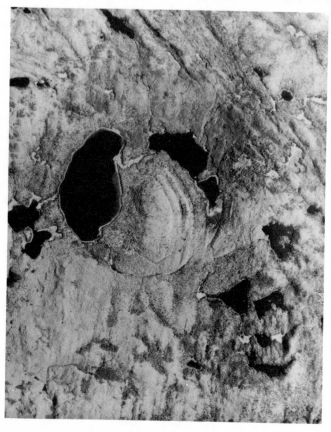

Figure 37.2 *Aerial photograph of the Brent crater in Ontario, formed 450 million years ago and originally 4 km in diameter. The original rim has been removed by erosion and the crater is partly filled by sediments*

plane would be barely off the ground by the time the Apollo body had passed over Montreal and hurtled beyond.

The energies involved in a major impact event are equally difficult to appreciate. A body about 200 m across, hitting the Earth at these cosmic speeds, releases an amount of energy equivalent to the explosion of about 100 million tons of TNT. This small body has enough energy to produce a crater about the size of the 4-km Brent crater in Algonquin Park, Ontario (Figure 37.2).

Large-scale impact cratering is usually considered to be outside the realm of traditional geologic processes. Unlike erosion, sedimentation, volcanism and to a lesser extent tectonism, huge impacts cannot be observed directly. It is also not particularly amenable to experiment! Even the detonation of nuclear devices scarcely approaches the energies involved in relatively small impacts. Planetary images provide important information on the form of craters but little on the geologic relationships. It is here that the terrestrial impact craters are an important source of data. Geologic and geophysical studies at terrestrial craters, such as the 23 impact craters that geologists now recognise in Canada, provide far more information on the impact process and its effects than the small number of craters might suggest. By combining the results of terrestrial and planetary studies with experimental and theoretical considerations, we now have a fairly clear picture of what happens during an impact. When a body strikes the Earth at 25 km/s, it penetrates into the ground about the equivalent of the body's diameter. As the projectile penetrates, and is being slowed down, most of the vast quantity of energy stored in the projectile by virtue of its cosmic velocity is transferred to the ground. Energy transfer is by means of an outward moving shock wave. The shock wave compresses the rocks and at the same time accelerates them away from the point of impact. At the point of impact, shock pressures are measured in megabars, millions of times normal atmospheric pressure, and initial radial accelerations are kilometres per second. A series of release waves follow the shock wave and bring the rocks back to normal pressures.

When the rocks decompress, not all the energy pumped into them by the shock wave is recovered. This excess energy appears as heat, and at large impacts the projectile and the rocks close to the point of impact are vaporised and/or melted. Initially, these impact-melted rocks can have temperatures of several thousand degrees Celsius – many times hotter than molten lavas that erupt

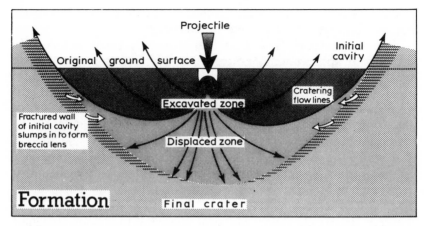

(a)

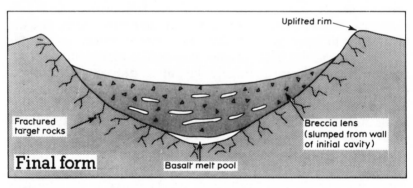

(b)

Figure 37.3 *Formation of a simple crater through the excavation and displacement of the rocks by the combined effects of the shock and release waves. The initial bowl-shaped cavity* (a) *is unstable and the walls collapse inwards, resulting in a final crater* (b) *partially filled by fragments (breccia) from the collapsed walls*

from volcanoes. Farther from the point of impact, the rocks remain unmelted but are permanently changed with peculiar features of deformation, such as shatter cones, microscopic dislocations, and solid-state glasses. These effects are collectively known as "shock metamorphic features" and are the principal means by which we can recognise ancient terrestrial craters.

The release waves also change the paths of some of the rocks accelerated by the shock wave. They deflect their movement to a more upward direction, which leads to the excavation and ejection of material. The amount of material ejected at the point of strike, or "target", is approximately 1000 times the mass of the projectile, and a cavity is formed partly by excavation and partly by displacement of the target rocks (Figure 37.3). For a crater a few kilometres across, the whole process of cavity formation takes place in a time span of about a minute. This can be compared with the time span of more common geologic processes that are measured in millions of years. Most of the material thrown out is deposited as a blanket of ejecta, which may extend several diameters beyond the crater. Some of the early ejecta, however, leaves at extremely high speeds and may be hurled into the upper reaches of the stratosphere where it can spread to even greater distances.

Figure 37.4   *The famous Meteor crater, Arizona, formed only a few tens of thousands of years ago and 1.2 km in diameter. It is virtually uneroded and is a smaller version of what the Brent crater (Figure 37.2) may have looked like shortly after its formation (Credit: Kenneth Fink, ARDEA)*

The initial cavity formed by excavation and displacement is about a third as deep as it is wide. At the Brent crater in Algonquin Park, the initial cavity is about 3 km across and over 1 km deep – as deep as parts of the Grand Canyon. This initial cavity is extremely unstable and the weakened and fractured walls slump inwards, partially filling the crater with broken rock or breccia and slightly enlarging the diameter of the final crater (Figure 37.3). Craters such as these are known as simple or bowl-shaped craters (Figure 37.4). Above some critical diameter, impact craters become what are known as complex structures. The critical diameter varies between planets and is a function of the nature of the target materials and the strength of planetary gravity. On Earth, Brent is the largest known simple crater in crystalline rocks, and at diameters greater than 4 to 5 km all terrestrial craters have a complex form. Complex structures are relatively shallow, with depth-diameter ratios of 1:10 or less, and are characterised by an uplifted central peak and/or rings. Apparently the displacements which go to form the initial, bowl-shaped cavity at simple craters are not locked in and the floor of the cavity rebounds upwards in complex structures to form a relatively shallow structure with a central peak and/or rings (Figure 37.5). Exactly how this rebound occurs remains the least understood part of cratering studies. Various lines of evidence point towards the rocks having little or no strength during rebound. This lack of strength and the similarities of the peaks and rings to what is seen when pebbles are dropped in water has led to the use of the phrase "hydrodynamic behaviour" to describe what happens during uplift.

Because of the small number of known terrestrial impact craters, one could argue that impact is a relatively minor geologic process on Earth. There is little doubt that it has profound effects on the local geology in and around the impact site. The definition of what is local, however, depends on perspective. The 210-million-year-old Manicouagan impact structure in Quebec represents the equivalent of 100–1000 times the energy of all the earthquakes occurring in one year, concentrated at one spot on the Earth's surface and released in minutes. A crater some 75 to 100 km in diameter was formed and rocks originally several kilometres below the surface were lifted up to form a central peak, which today stands some 500 m above the ground. As much as 1000 km^3 of the target rocks were melted: the effects of the impact are still visible in the rocks over an area of 20 000 km^2

There is some evidence that relatively recent, large, impact

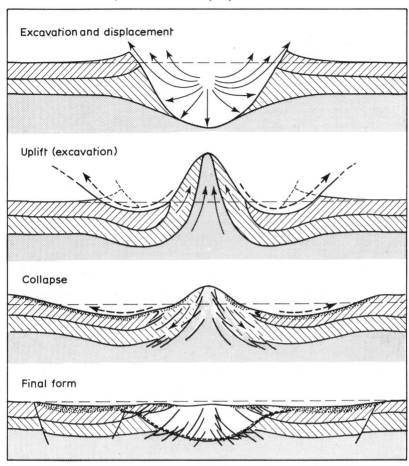

Excavation and displacement

Uplift (excavation)

Collapse

Final form

Figure 37.5　*The initial stage of formation of complex crater structure is similar to that in simple craters (Figure 37.3), but early displacements are not locked in and the crater floor rebounds, then collapses, resulting in a shallow final crater with a small depth to diameter ratio and an uplifted central region*

events may have also affected the Earth on a global scale. Geochemical data indicate a major accumulation of extraterrestial material, including iridium and other heavy metals typical of meteorites but not of the Earth's crust, in sedimentary layers marking the boundary between the Cretaceous and Tertiary periods, 65 million years ago. This boundary correlates with the

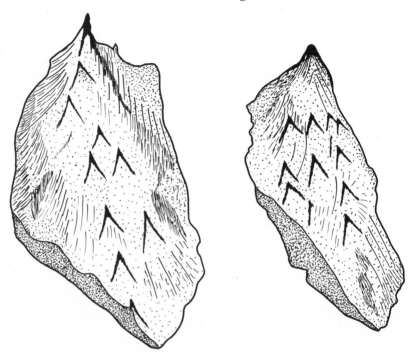

Figure 37.6   *Many geologists believe that shatter cones are formed by meteoritic impacts. They occur in many types of rock and vary in length from a few millimetres to metres*

mass extinction of approximately half the faunal genera living on Earth at the time, and it has been suggested that a major impact event produced a sudden climatic change that led to the mass deaths of many organisms, including the once dominant dinosaurs.

Further evidence for the possible influence of impact on terrestrial life has been unearthed at the boundry between the Oligocene and Eocene epochs 38 million years ago. A similar extraterretrial signature occurs at this boundary, which also coincides with the disappearance of 70 per cent of the microscopic marine organism *Radiolaria* living in tropical seas, and major changes in the mammalian and floral populations of the Northern Hemisphere. No very large impact crater has been found which corresponds in age with the Cretaceous–Tertiary boundary. It is generally presumed, therefore, that the impact occurred in the ocean, a suggestion supported by some of the geochemical data. At the boundary between the Oligocene and Eocene epochs, however, there is a possible candidate: the Popigai impact structure in Siberia, which is about 100 km in diameter and somewhere between 34 and 40 million years old. Much work remains, however, before this suggested link between impact and global biological changes can be considered as a firm working hypothesis.

Although no evidence is preserved, the Earth must have been subjected to the early high flux of large bodies that produced the 1000 km-sized impact basins on the other planets. Impact events of these magnitudes are truly global in their effects. For example, the event that produced the 1300 km Mare Imbrium basin on the Moon released up to a million times more energy than is released by the Earth in a year. It excavated the crust to a depth of 50 km,

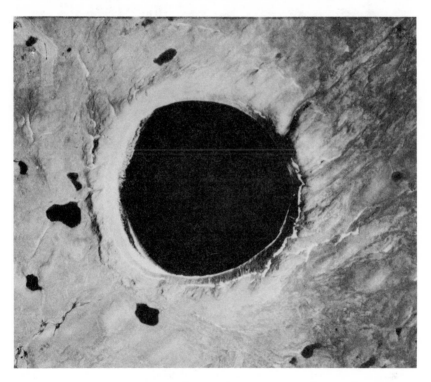

Figure 37.7 *Two ages of Canadian Craters: the 3.2-km one at New Quebec (above) was created 10 million years ago, while the vast 75-km wide Manicouagan crater (overleaf), viewed here from a satellite, was formed 210 million years ago*

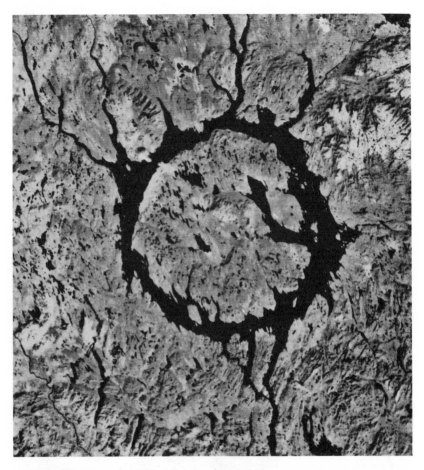

Figure 37.7   *(cont.)*

ejected 10 million km^3 of material and melted as much rock as is
contained in the Columbia River Basalts, which cover much of the
states of Washington and Oregon.

In addition to immediate effects, the formation of basins results
in large, deepseated thermal anomalies, which for the Moon
persisted for up to a billion years. The effects of such events on the
early history of Earth are not known. There may originally have
been 25 or more impact basins larger than a 1000 km across on
the early Earth, and the events which formed the basins may have
helped to trigger the basic differences between oceanic and
continental crust. Whatever the case, recent discoveries have

raised impact cratering from a minor scientific curiosity to a fundamental geologic process in the early history of the planets. It seems that cratering may have exerted some control over the evolution of the Earth and the life that exists on it.

17 November, 1983

38

Controversy over Earth's interior

What lies deep inside the Earth? It is one of the frustrations of geology that in many ways we know more about the composition of distant stars than we do about the composition of our own planet. At least we can see light from other stars and analyse it by spectroscopy; the Earth's interior can be probed only by waves generated by earthquakes or large explosions. With such observations as a guide to density, and guessing that our planet is made of material similar to that in meteorites, the established view is that the Earth's core is composed mainly of nickel-iron. But J. M. Herndon, of the University of California, San Diego, argues against this view.

Herndon also assumes that the Earth must have been built originally from meteoritic material. But there are two very different kinds of meteorite: those rich in nickel-iron, and the so-called chondrites rich in silicates. What if the inner core of our planet is not a nickel-iron mix, but nickel silicide? As this is the only sensible alternative to the nickel-iron model, it is worth looking at the implications. And as just about the only direct observational data regarding the deep Earth give us density estimates, Herndon has compared the mass density ratios for the core of the Earth and the lower mantle (the region between the core and the upper mantle and crust) with those of the components of the chondrite meteorites. He finds that the mass ratio for the two components of the Earth is 1:49, which compares well with the ratio 1:43 for the silicates and sulphur-based iron alloy components of chondrites (*Proceedings of the Royal Society*, vol. A372, p. 149).

There are no signs yet of planetary geologists and astronomers beating a path to Herndon's door to acclaim this discovery as a great new truth. But the theorists are going to have to find *some* way of modifying their models to take account of the observational evidence.

"MONITOR", 9 October, 1980

PART FIVE

The Payoffs

39

Prospecting for oil in the North Sea

TOM GASKELL

This month seismic exploration begins, over a large area of the sea lying between the oilfields of the English Midlands and those of Holland and Germany, in the hope of detecting formations in the underlying strata which might indicate the presence of oil or natural gas.

The announcement last week that Shell, BP and Esso are to carry out a joint seismic survey of a large area of the North Sea may prompt the question: Why is it worth while looking for oil in an area which is covered by so many fathoms of water? In fact this deployment of the exploration resources of the companies is a logical development of work that has been going on for the past 25 years.

In 1936, BP (at that time known as the D'Arcy Exploration Company) started a search for oil in Britain. Gas was soon found at Cousland near Edinburgh and at Eskdale near Whitby. In 1939 Eakring, the first of several oilfields in the Nottingham area, was discovered in time to make a useful contribution during the war, when shipping to carry oil from abroad was at a premium. Although the production of oil from the Midlands accounts for only a small fraction of present-day demands, it does suffice to show that there are oil-bearing rocks in the sedimentary layers that comprise the geological picture of North-West Europe.

The Eakring oil is contained underground in layers of sandstone which are situated geologically in the Coal Measures in the Carboniferous series. Confirmation that the north-west Europe sedimentary basin is a potential oil-producing area is provided by the discovery of fields such as Schoonebeek in Holland and of many producing horizons in the Jurassic and Cretaceous rocks

(150 million to 120 million years old) in Germany. In fact German production is now running at about 7 million tons a year.

It is possible, by comparing the rock strata encountered in the Midlands with their counterpart in Holland and Germany, to draw a picture of the shallow seas and land areas of north-west Europe during those geological periods when the rocks containing the English, Dutch and German oilfields were being laid down. From Carboniferous times, the North Sea between Holland and the English Midlands has probably always been an area of shallow water in which has been accumulating the silt and sand brought down by the Rhine, Elbe and other great rivers of Europe (see Figure 39.1). The dip of the strata in East England and in Holland indicate that the centre line of the depositional basin is in the North Sea, and this supposition has been confirmed by experiments made by Mr A. H. Stride of the National Institute of Oceanography in conjunction with Professor Maurice Ewing of Lamont Geological Observatory. They carried out seismic refraction experiments (in these, the source of the pressure waves and the seismometer which detects them are separated by a considerable distance, as in the study of earthquakes) and, although they did not succeed in determining the total depth of the sedimentary basin, they did show at least 11 000 feet of sedimentary material which they considered to be made up of 3000 feet of clay and chalk and over 8000 feet of limestone or sandstone.

We might ask why no one has thought of looking for oil before in this large chunk of sediments. Part of the answer is supplied by this quotation from an editorial in the *New Scientist* of 30 January, 1958:

"The United Nations Commission on Offshore Rights should decide in the next few weeks what are to be the rules for ownership of minerals found on continental shelves. This may give a fillip to oil exploration in the North Sea. Already German activity is reported in the Baltic; the Dutch are drilling almost on the sea-shore, and Esso have taken over the old Gulf prospecting rights in Denmark. It is probable that ownership will be determined by nearness to a particular country, so that countries with bulges convex to the sea, like England and Holland, will tend to have some advantage."

The matter is still in a state of suspense. A Convention was duly drawn up in April, 1958, which gave the mineral rights of the continental shelves to the adjacent nations, subject to demarcation

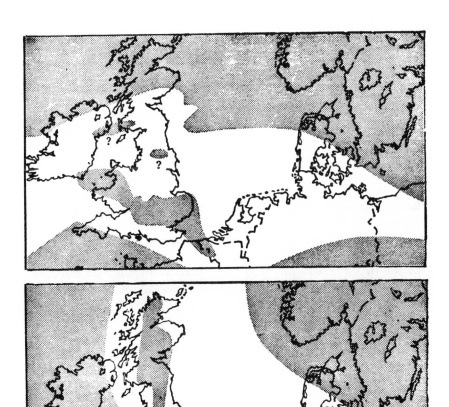

Figure 39.1 *Sea and land in North-West Europe (top) during Carboniferous times (about 150 million years ago) and (bottom) during Jurassic times (about 30 million years later). Land areas are tinted (adapted from J. L. Wills, A Palaeographic Atlas of the British Isles and adjacent parts of Europe (1951))*

between them. It has not come into force because enough
countries have not yet ratified it.

Again, although oil was found in Holland and England, long
years of search in Denmark failed to locate any productive fields.
That clearly had a damping effect on those who were hoping for
big discoveries offshore. Another reason why the offshore prospecting has been held in abeyance is that it is very expensive to
drill oil wells at sea; to find and to produce small oilfields the size
of those in Nottinghamshire becomes uneconomic when they are
covered by water. However, just as the first offshore drilling in the
Gulf of Mexico was an extension from the successful swamp
drilling in Louisiana, so in time the oil men will inevitably move
out from Holland and England into the North Sea.

The effort that is now being applied to offshore drilling has
shown ways of making production at sea much less expensive. For
example, drilling ships appear to be much cheaper to operate than
the giant platforms which are raised on legs from the sea bed,
while placing the valves and fittings of the completed oil well on
the sea bed obviates the need for expensive production towers.

In addition to the technical improvements, there have been some
new finds which make the North Sea picture much more optimistic. New natural gas wells in Holland are reported to be large
producers, and gas these days is in many ways as useful a
commodity as oil. Thoughts naturally turn to Lacq, the enormous
gas field in the foothills of the Pyrenees. If such a field can exist in
the South of France, then there may be a similar one waiting to be
discovered under the North Sea.

The exploration programme which starts this month will consist
of probing of the rock layers of the sea bed by means of seismic
waves produced by explosions (Figure 39.2): in these seismic
reflection studies the technique resembles echo sounding and the
source of the waves and the detectors are close together (Figure
39.3). An enormous area will be covered during two seasons from
July to November and again in the summer of 1963 for another
four months (Figure 39.4). Seismic exploration can be conducted
much more rapidly at sea than on land, for two reasons. In the first
place, on land holes must be drilled for the explosive so that plenty
of energy is transmitted to the ground. This shot-hole drilling
slows up the operation, but at sea it is necessary only to trail the
charges behind the firing boat and detonate them in the water.
Secondly, the instruments which record the seismic waves as they
return to the surface after being reflected from the rock layers

Figure 39.2   *A plume from an explosion. The shot is fired within a few feet of the surface: more seismic energy is produced by a deeper explosion, but confusing impulses originate in the oscillation of the gas bubble that is created* (Credit: The British Petroleum plc)

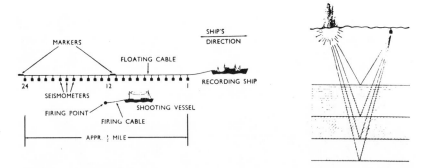

Figure 39.3 *Sound waves travel out in all directions from the explosion. Some are reflected from the sea bed, others from deeper horizons where one layer of sediment rests upon another*

beneath the sea bed are towed in a line behind the ship, instead of being laboriously laid out on the ground. The position when the explosion occurs is noted on board the ship by means of the Decca Navigator system. Therefore, although running a ship is expensive, seismic reflection geophysics can be carried out so speedily at sea that the overall operation costs no more than it does on land.

The object of the reflection work is to find whether the geology beneath the sea bed shows formations which could be suitable reservoirs for oil or gas. It is not, of course, possible to tell without drilling whether oil or gas is actually present – promising underground traps often disappoint the oil prospector by turning out to contain only water. It is hoped that some large hump-like formations may exist beneath the shallower sedimentary layers in the North Sea, because this type of anticlinal structure can be associated with really extensive reservoirs. If some promising structures are located, the next step will be the expensive one of proving what they contain, by drilling into them.

It is possible, too, that a continuation will be made of Professor Ewing's refraction work, because this could delineate the base of the sedimentary basins. The decision to conduct the survey will be warmly welcomed by geologists, because whatever the yield of oil, a great deal of new and interesting information is certain to emerge concerning the geological history of north-west Europe.

5 July, 1962

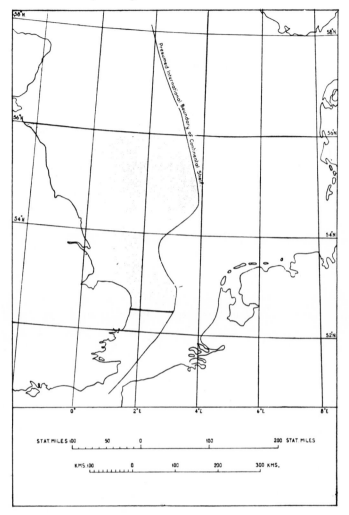

Figure 39.4   *The approximate area of the North Sea which will be under investigation during the next two years*

40

Fire and water

PETER FRANCIS

The theory of plate tectonics has revolutionised concepts of ore genesis. Modern research shows that most metalliferous ore bodies result from the interaction of young igneous rocks with large volumes of water.

"Many persons hold the opinion that the metal industries are fortuitous and that the occupation is one of sordid toil, and altogether a kind of business requiring not so much skill as labour. But for myself . . . it appears to be far otherwise." Thus wrote Georgius Agricola, the founder of scientific studies of metal mining in *De re Metallica,* published 1556. Those who still think that the mining industry is one of "sordid toil" – and there are many – would have had their eyes opened at a recent meeting in London. Jointly organised by the Institute of Mining and Metallurgy and the Geological Society of London, the purpose of the meeting was to explore new ideas about the relationship between volcanic processes and the formation of ore bodies.

Almost as soon as it had been demonstrated that all the world's volcanic and seismic belts could be fitted together into the global pattern of plate tectonics, geologists realised that the formation of ore bodies was bound up with the same pattern. According to plate tectonics, the Earth's crust is divided up into major plates, which are separated from one another by *constructive* and *destructive* plate margins. Once these had been identified all over the world, it became pretty obvious that many important ore deposits are located on either constructive or destructive plate margins, or on ancient remnants of them.

Constructive plate margins are synonymous with mid-ocean ridges and, since these are mostly below sea level, it is difficult to get a look at the ore bodies associated with active ones. (But not

Figure 40.1 *San Pedro and San Pablo volcanoes in the Atacama desert of North Chile. These 6000-metre-high andesite volcanoes may be sitting on top of porphyry copper deposits, if the model illustrated in Figure 40.3 is correct. San Pedro (left) has an active fumarole which is currently depositing sulphur (Credit: Dr Peter Francis)*

impossible, as we shall see.) Due to all the stirrings and shiftings of the crustal plates that constitute continental drift, there are many places around the world where great slabs of oceanic crust have been literally shoved up on to dry land. Many of them contain rich ore bodies. The rocks which make up these stranded slabs are known as *ophiolites*.

The best place to see ophiolites in Europe is on the Troodos massif in Cyprus. Here, copper ores have been continuously exploited since the earliest historic days – the very name "Cyprus" is derived from the Greek word for copper. The copper ores occur as copper sulphides, mixed with some zinc sulphide, in massive deposits sandwiched between layers of basaltic "pillow" lavas. These lavas were erupted under water at a former mid-ocean ridge during the Mesozoic era (the long geological interval falling between 70 and 220 million years ago). Ore bodies associated with Troodos-like ophiolite rocks have also turned up in many other parts of the world from Norway to Newfoundland, and in rocks ranging from 60 to over 600 million years old.

Troodos type copper deposits, however, are overshadowed in economic importance by copper ores formed at the other end of the plate tectonic conveyor belt, the destructive plate margins. Many of the world's biggest copper mines are clustered around the shores of the Pacific Ocean, in particular along the western coasts of the Americas (Figure 40.2). They parallel the great belt of active andesite volcanoes that encircles the entire Pacific. This "Ring of Fire" marks the site of present-day destructive plate margin activity. Here, the Pacific ocean crust is forced down into the mantle along so-called Benioff zones, to the accompaniment of frequent earthquakes and rather less frequent volcanic eruptions.

The western American copper deposits are all of a single type

Figure 40.2 *Porphyry copper deposits in the western Americas. Only those that are actively being mined are shown; many more deposits certainly exist*

known as "porphyry coppers". These were produced when intrusive masses of molten rocks (approaching granite in composition) rose towards the surface from the Benioff zone where they were generated. Some of them were probably capped by andesite volcanoes. Many separate intrusions – injections of molten rock into the overlying strata – took place over a long period to give rise to a single ore body. Some intrusions were barren; others seem to have been barren to start with but were subsequently mineralised by hot solutions carrying copper, gold and molybdenum. The end result was the formation of huge masses of rock which generally have a low grade of mineralisation (less than 1 per cent copper) but which together make up most of the world's reserves of copper.

A rather different situation prevails on the opposite shores of the Pacific. Here the interaction of oceanic plates along the destructive plate margin has given rise to the chains of island arcs which start with Japan and extend southwards, disappearing eventually in the tangle of islands that is Melanesia. Some porphyry coppers are found there but two different kinds of ore deposit have been recognised in the Japanese islands, the Kuroko and Besshi type. Both are massive sulphide deposits.

The Kuroko type are lead-zinc-copper-silver sulphide ores which were produced by submarine eruptions of rather acid volcanic rocks (the term "acid" refers to the silica content of the rocks), which took place rather late in the evolution of the island arc, and well away from the front of the arc. Besshi ores are mostly copper-iron sulphides. These, too, seem to have been produced by submarine volcanic activity, but this time of more basaltic (less acid) rocks at greater depths and earlier in the evolution of the arc system.

It is becoming clear that the zone behind the active front of an island arc is of prime importance in the formation of ore deposits. Interestingly enough, speakers suggested at the London conference that the Troodos type ores, formed at constructive plate margins, were probably not formed at major mid-ocean ridges such as the mid-Atlantic ridge, but rather at smaller spreading centres in the marginal seas that lie between island arcs and larger plates. (The Sea of Japan is a modern example of such a marginal sea.) The clear links between ore genesis, volcanic activity and plate tectonics have not been lost on economic geologists, many of who have been as keen as Agricola to demonstrate that the search for minerals, at least, is anything but "a kind of business requiring not so much skill as labour". An early example of the application of plate tectonic thinking to mineral prospecting was the discovery of the huge Bougainville copper deposit – a porphyry copper – in the Solomon islands.

The role of the sea

It is reassuring, in these days of increasing concern over the rapid rate of depletion of the Earth's resources, to know that the plate tectonic processes responsible for bringing metals up to the surface from the mantle, redistributing, and concentrating them, are still going strong at the present day, albeit at rates much slower than they are being extracted by man. Some fascinating oceanographic work in the Red Sea has shown the presence, along the axis of the supposedly embryonic mid-ocean ridge there, of pools of hot mineral-rich brines trapped in basins known as "deeps". The brines are of hydrothermal origin, apparently having been exhaled from submarine fissures along the ridge. Interaction between the brines and ordinary sea water caused precipitation of sulphide and silicate minerals, so the sediments around the deeps are rich in zinc, lead and copper. The Atlantis II Deep contains some three million tons of zinc, one million tons each of copper and lead, and perhaps 5000 tons of silver within an area of 50 square miles. Both Sudan and Saudia Arabia are eyeing these deposits covetously, and trying to decide how to get at them.

Even bigger copper deposits are believed to be present in the sediments on the flanks of the East Pacific Rise. It is possible that copper-rich rocks produced in this way at mid-ocean ridges may also have been responsible for the circum-Pacific porphyry copper ores, after being transported for thousands of kilometres on the plate tectonic conveyor belt and then carried down into the stewpot of the Benioff zone. Partial melting of the oceanic crust takes place there, so some of the copper may find its way upwards along with the silicate magmas. The porphyry coppers now being mined are all at least 10 million years old, but the active volcanoes that now surround the Pacific may be sitting on top of younger ore bodies, not yet exposed by erosion. Many of these volcanoes are in a fumarolic state – not aggressively active but gently exhaling warm breaths of sulphurous steam. The deposits around the fumaroles often contain traces of copper sulphides which give an inkling of mineralisation processes going on beneath (Figure 40.3).

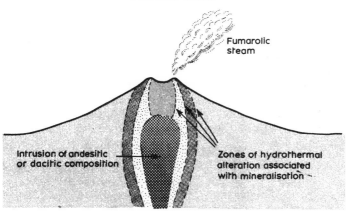

Figure 40.3 *One possible model for the relationship between andesite volcanoes and porphyry copper deposits. The volcano and the intrusion beneath it are linked in circulatory hydrothermal solutions. Extensive alteration is observed in the cores of old volcanoes, and in the rocks around porphyry copper intrusions (after Sillitoe, Halls and Grant,* Economic Geology, *1975)*

Some scientists discount the oceanic crust as a source of ore metals, suggesting instead that they may be derived from the lower continental crust or the upper mantle. Wherever they come from, the background levels of the useful metals in their host rocks are undoubtedly very low. Very intensive processes are required to concentrate them into economically workable ore bodies. Much of the interest at the London conference was centred on the mechanisms which bring about such concentrations.

Geochemists have known for a long time that circulating hydrothermal waters – or juices, as Agricola would have called them – are capable of leaching out valuable metals from large volumes of rocks and redepositing them in smaller, much more concentrated ore bodies. But it is only now that geologists are realising the colossal scale on which this leaching process takes place. It is becoming apparent that the interaction between the waters of the oceans and the rocks of the oceanic crust is one of the most important processes in geology.

This realisation is the result of many sophisticated geochemical studies, but most importantly the application of stable isotope studies to ore bodies and volcanic rocks. Hydrogen, oxygen and

sulphur have light, stable isotopes which have provided geochemists with a very powerful tool. During natural geological processes,
light and heavy isotopes respond slightly differently to the
conditions, and the stable ones undergo fractionation. Through
geologic times, the consequences of this fractionation have become
very pronounced. For example, the hydrogen and oxygen isotope
ratios in primary magmatic waters (believed to have come direct
from the mantle) fall within a very narrow range, no matter where
in the world samples are collected. This range is quite different
from that found in sea water, which again shows identical ratios
all over the world. Rain water shows distinct ratios again, though
these do vary from place to place. The stable isotope ratios, then,
provide a means of fingerprinting the water found in association
with rocks and ore bodies, and enable its origin to be determined
with some confidence.

An important discovery has emerged from these fingerprinting
studies. The hydrothermal juices responsible for the formation of
most ore bodies did *not* simply come up from the mantle along
with the igneous rocks involved, such as the pillow lavas in the

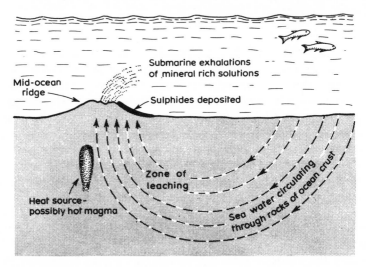

Figure 40.4   *Heat sources beneath mid-ocean ridges power
large convection cells which draw sea water through the
rocks of the ocean crust, where they leach out valuable
metals. They then exhale through vents along the mid-ocean
ridge where the metals are redeposited in economically
significant concentrations*

case of the Troodos sulphide deposits. Rather, isotope studies reveal, huge volumes of "ordinary" sea water soaked their way deep down into the rocks around the ore body, and were then carried back to the surface by the convection cell set up over the hot spot that was originally responsible for the volcanic activity (Figure 40.4).

In its passage through the rocks, the sea water would react with them chemically. Chlorine from the sea water would help in the leaching process, forming soluble metal chloride complexes. When they reached the sea floor, these hot, metal-rich solutions would react with the cold sea water and, in appropriate circumstances, sulphides would be precipitated, just as is happening in the Red Sea today.

The scale on which this process must take place can scarcely be over-emphasised. Geologists believe that hydrothermal solutions may percolate down as deep as 10 kilometres into the rocks of the ocean crust. Ten kilometres thickness of fresh, unleached crust would contain a large amount of most valuable metals distributed uniformly at very low concentrations. So it is not difficult to see how economically useful amounts of them can be accumulated by such thorough flushing processes.

As this profound new realisation demonstrates, we have come a long way since the days of Agricola. He would certainly have been delighted by the advances in mineral sciences revealed at the conference. For, as he said, " . . a miner must have the greatest skill in his work, so that he may know first of all what mountain or hill, what valley or plain can be prospected most profitably, or what he should leave alone . . . then he must be familiar with the many and varied species of earths, juices, gems, stones, marbles, metals and compounds . . ." Those requirements for mineral scientists are as important now as they were 400 years ago.

12 February, 1976

41

Sea-bed geology goes up the mountains

SEAN McCUTCHEON

By drilling into the mountains on Cyprus geologists are hoping to learn more about the sea bed. The highest rocks on the island, it seems, once lay deep beneath the sea.

Early one evening last spring in the Troodos mountains of Cyprus a peasant was climbing up a winding road, driving a donkey laden with goat fodder. As he passed our party he saluted with polite astonishment. Some of the group were hammering at the rocks through which the road cut. Pacing and gesticulating, the others were debating whether or not the flat patch by the road could be widened to accommodate a drill rig by blasting the cliff. The peasant was witnessing members of an informal group of Earth scientists known as the International Crustal Research Drilling Group (ICRDG) during the early stages of an intriguing and seemingly paradoxical research venture: probing into a mountain range to learn a good deal about the ocean floor.

One member of the ICRDG is Ian Gass, a geologist who now teaches at the Open University. During the 1950s, when he was a member of the Cyprus Geological Survey, Gass mapped the Troodos massif and worried about its origin. It was, he decided in the end, an ophiolite. Ophiolite comes from *ophis*, Greek for "snake", and it originally signified rocks, such as mottled-green serpentinite, that resemble snakeskin. It implied nothing about how such rocks were formed. "The 19th century geologists", according to Gass, "would draw a line around anything green and dirty, call it an ophiolite, and walk away." Early this century, however, the meaning of the word was broadened to include the other rocks such as "pillow" lavas, commonly found with serpentinite, although until recently there was no generally accepted explanation for this association of rocks.

Figure 41.1 *This rock formed in layers as minerals crystallised out from the molten rock of a magma chamber beneath the sea (Credit: Professor F. Vine)*

After a detailed study of the Troodos massif, Gass became convinced that all the 3000 sq km was a fragment of ocean floor that by some monumental feat of Earth moving had been stranded on Cyprus. He co-authored a paper in which he identified these mountains, which cover a quarter of the island, as an ophiolite, using the word in its modern sense. Ophiolites, geologists now believe, are land-bound fragments of oceanic crust. Several hundred have been identified, and many are being studied for the evidence they can yield on how the oceans form.

How do the oceans form? The first version of the currently-accepted answer to this question was developed in 1960 by Harry Hess of Princeton University. Hess commanded a troop ship during the Second World War and the vessel's echo sounder had traced out curious mountains on the floor of the Pacific. Speculations on their origins, in what he called "an essay in geopoetry", led Hess to outline the concept of sea-floor spreading, the gist of which is as follows. At ridges in the mid-oceans molten rock rises from the mantle below the crust and spreads out on either side to form new ocean floor. The new oceanic crust welds to the edges of two plates (huge pieces of crust that move slowly over the mantle beneath). This new oceanic crust migrates from its place of birth, eventually plunging to destruction back down in the mantle, when the oceanic plate collides with a continental plate.

When new and still unorthodox, Hess's concept fired the imaginations of a number of the Earth scientists who were later to get together on Cyprus. Fred Vine was one of these. In 1963 Vine had just graduated in geophysics from Cambridge. With Drummond Matthews, his supervisor, he showed that the sea-floor spreading combined with periodic reversals in the direction of the Earth's magnetic field could neatly account for the striking pattern of magnetism on the flanks of mid-ocean ridges.

This work, in time, helped move geological opinion from scepticism to belief in sea-floor spreading and in the overall concept of plate tectonics. One reason for resistance to the new paradigm was the sheer impossibility of convincing geologists in the traditional way: by convening on an outcrop and hammering

Figure 41.2 *Bulbous shapes of pillow lava formed as hot lava erupted to be quickly chilled in cold sea water (Credit: Professor F. Vine)*

Figure 41.3 *Columns of rock formed as lava erupted onto the cold ocean floor (Credit: Jim Hall)*

out the evidence. Though they cover more than 70 per cent of the Earth's surface, the rocks of the sea floor are virtually inaccessible. Between geologist and geology there lies a daunting barrier: the deep and rolling ocean.

Jim Hall, a founder of the ICRDG and one of its driving forces, was doing geology on land in his native Britain and in Africa, when he was, like Vine, "bitten by the bug of the ocean floor". In 1971 he moved to Canada, to Dalhousie University and that same year sailed out of Halifax on board the research ship Hudson for his first field trip to study marine geology. He gathered rocks, but found a dredge such as the Hudson's to be a crude sampling tool. There was no way to pinpoint where on the sea bed the dredged rocks had come from, nor any way to penetrate into the ocean's basement. For good data, Hall decided, he would have to drill.

Hall's office at Dalhousie is decorated with souvenirs of the drilling expeditions he has since helped organise. With co-workers he probed into the floor of the Atlantic from the margins of San Miguel in the Azores and Bermuda. In 1974 he was on board the *Glomar Challenger*, for Leg 37 of the Deep Sea Drilling Project. On that cruise the ship's drilling bit, after being lowered through

3 km of water, bored through a thin veneer of sediments and 600 m into the underlying basalts; it had successfully penetrated the ocean crust for the first time.

On the *Glomar Challenger*, and back on shore, Hall met others who shared his enthusiasm for the ocean floor. They talked about pooling their resources. By 1978 they had formed the International Crustal Research Drilling Group and a team of a dozen or so, along with graduate students and drillers, was at work in Iceland. The island is an ideal location where new ocean crust wells up at active ridges near the surface, rather than deep beneath the ocean. But this very difference means Icelandic crust is not the same as normal ocean crust.

Early in 1982 the headquarters of the ICRDG moved from Hall's office in Dalhousie University to Cyprus. When I visited Hall and his colleagues there they were beginning their most ambitious project yet. The team is truly international and includes: Fred Vine from England; Paul Robinson, director of the Cyprus Project, and Paul Johnson from the US; Hans-Ulrich Schmincke from West Germany; Kent Brooks from Denmark; Ingvar-Birgir

Figure 41.4 *Drilling on the upper pillow lavas on the Troodos massif (Credit: Jim Hall)*

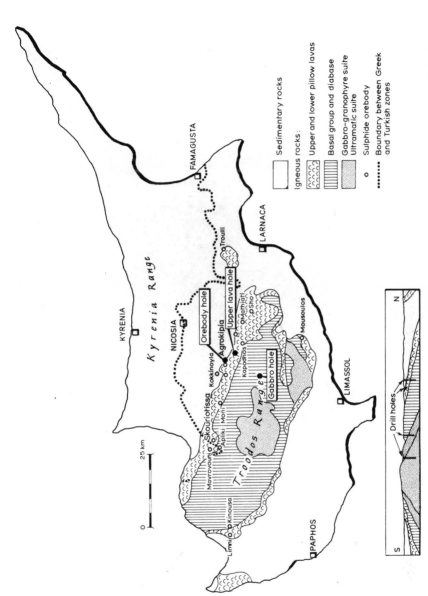

Figure 41.5   *The Troodos massif appears to be a fragment of ocean crust. Drill holes, as shown, will sample different layers that once lay under the sea floor*

Fridliefsson from Iceland; and Andreas Panayiotou of the Cyprus
Geological Survey and George Constantinou, its director.

The ICRDG is a collective; its members pool their skills and
their grants. But there are difficulties in finding money for research
drilling. In Cyprus the plan was to obtain a total of 5 km of core
from separate holes sampling the upper 4 km of the ocean crust.
For this the drilling costs will eventually total an estimated
Canadian $1.1 million (nearly £600 000); and most of this has, to
date, been raised. There is, it seems, a prejudice against research
drilling. The geologists sense that some funding agencies

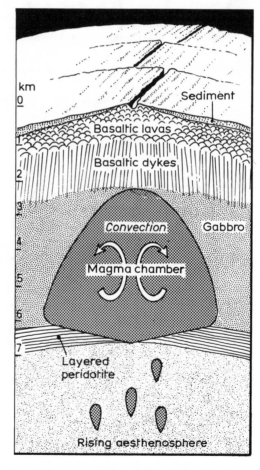

Figure 41.6   *Ocean crust forms in different layers when hot
rock cools as it nears the sea floor*

"obviously feel that ophiolite work is a waste of time. They think you can't learn anything about the oceans if you don't get your feet wet."

From limited evidence, from drill cores, from seismic data, and from what has been learned by field work on ophiolites, Earth scientists have put together a composite picture of the layered oceanic crust (Figure 41.6). At the lowest level are mantle rocks such as serpentinite and harzburgite. These are the congealed and metamorphosed residue left in the bowels from which magma, or molten rock, has been vomited upwards.

In the middle layer is greenish gabbro, molten rock that never made it to the surface of the crust. Instead it slowly froze in a magma chamber into a mush of coarse crystals. The upper layer is of lava. Where the rising lava froze like playing cards in the fissures of a spreading sea-floor are the so-called sheeted dykes, vertical slices of fine-grained lava. Above, where hot lava erupted from the crust to be chilled by cold sea-water are purple, bulbous pillow lavas.

Drilling brings to light a vertical column of rocks in which history is written albeit tersely, in the cryptic shorthand of rock sequence, texture, composition. The sequence above, starting with the lavas, is what geologists would expect to find in a hole piercing 6 km into the ocean crust. But on Cyprus a single hole is impractical; cost and difficulties mount rapidly with depth, so instead a number of holes (probably five) will be drilled in Cyprus. Because the Troodos ophiolite has been uplifted and eroded, a slice through successive layers of oceanic crust has been exposed, high and dry on the main hillsides of Cyprus. Thus the five holes will sample, in overlapping segments, the uppermost 4 km of the ophiolite (see map). These holes will probe the vertical relations between types of rock whose horizontal relations nature has already revealed in creating the Troodos massif.

At Mount Olympus, the highest point of the massif, mantle rock is visible. The slopes around are gabbro. (Gabbro weathers readily, forming the good soil on which grow the grapes that go into the sweet wines of Cyprus.) In a ring around the gabbro are the sheeted dykes and in the outermost ring, marking the periphery of the massif are the pillow lavas. In climbing down from Mount Olympus and passing over these concentric circles, you can imagine you are tracing the path of magma rising at the mid-ocean ridges, eventually to spew through vents, producing large quantities of mineral ore. Of all the phenomena submersibles

Figure 41.7   *The Troodos massif with Mount Olympus in the background (Credit: Professor F. Vine)*

have observed on the ocean's floors, the strangest are probably the black smokers – the hot springs on the East Pacific Rise that spew forth clouds of sulphide minerals (*New Scientist*, vol. 92, p. 7). To see a black smoker is to see a body of ore actually being formed. To drill beneath an ore body, as the ICRDG has done in Cyprus, is to study the plumbing of a fossilised black smoker.

As I write, the drilling phase of the Cyprus project is still underway, with four holes completed. The first, which sampled the upper part of the extrusive lava layer, had to be abandoned at 500 m because of technical problems. A second and a third hole have been drilled beneath the deposits at Agrokipia. All the cores obtained have been described in detail, and samples taken for geochemical, paleomagnetic and other laboratory studies. The fourth hole has reached a depth of 1850 m beginning at the base of the sheeted dykes, so as to sample the gabbro. Its purpose is to gather sufficient information to answer questions about magma chambers in oceanic crust. If funding permits, a fifth hole will be drilled to complete the sampling of the upper extrusives.

What can be learned from such research? Previous work by the ICRDG has shown, for example, that the growth of the sea floor is discontinuous. Eruptions on the Mid-Atlantic Ridge occur sporadically, each major one fed by a distinct magma chamber. The group also has evidence, from the Azores, of volcanoes rising above the waves only to sink again and again beneath the weight of accumulating lava. But, says Hall modestly and realistically, "Drilling has taught us that the first models we worked with were only the simplest of a large number of possibilities that were consistent with the observations. The information we are gaining is essential in developing more realistic models."

The project on Cyprus will shed light on the origin of the Troodos ophiolite, and on the structure of oceanic crust. By revealing the mechanisms of the formations of ore bodies on the ocean floor, it may lead to new techniques for the prospecting of minerals.

Curiously though, in Cyprus finding water is the paramount geological task: the growth of agriculture and hotel development added together have helped to drain the aquifers. The lakes, coloured blue on maps of the island, are dry. In every river there is a dam, and almost all the considerable revenue from tourism is spent on water projects. But the shattered gabbro of the Troodos ophiolite may well hold water, and information gained by drilling into it may be of great practical value on this semi-arid island.

But geology, is not done just to be applied. Even now, when it has in plate tectonics a unifying, fundamental theory, much of it remains an observational science, intuitive to the point of subjectivity. Doing geology is like playing outdoors with enigmatic jig-saw puzzles: giant puzzles in the two dimensions of the Earth's surface, the third dimension of a drill hole and the fourth dimension of time. There are always pieces missing; and ample room to use the imagination.

24 February, 1983

42

The world is a bit cracked!

JOHN NORMAN AND MUO CHUKWU-IKE

Satellite pictures of the world are now revealing huge crustal fractures showing evidence of its past history and possible locations of sources of minerals and thermal energy.

For several decades geologists have been looking down on the world with the aid of air photographs. One of the major advantages of this high viewpoint is the ease of detecting steeply inclined rock fractures, sometimes even where glaciers have moved material to cover them. Specialised studies of the fracture patterns are helping some photogeologists to detect concealed structures that might contain oil, water, or intrusive rocks related to economic minerals. A lesson from these studies has been that the scale of the photography used is closely related to the size of the features detected. Doubling the flying height of an aircraft halves the scale of photography, and each photograph then covers four times the area. In this greater area large features can be better sensed in their setting, so we began to look forward to the results of reducing the scale by two orders when US generosity made available the products of imaging devices scanning the Earth from satellites.

Literature on the new, bigger and better fractures (showing as lineaments) has grown into a torrent, but although many accompanying diagrams show patterns that cry out for interpretation, there are surprisingly few explanations that extend beyond local interest. We report here an interesting pattern from one area with our own very speculative interpretation. We hope unconvinced readers will produce practical alternatives for *New Scientist*'s correspondence columns.

Figure 42.1 shows one of a number of sets of parallel lineaments in Nigeria revealed by the multispectral scanner carried by

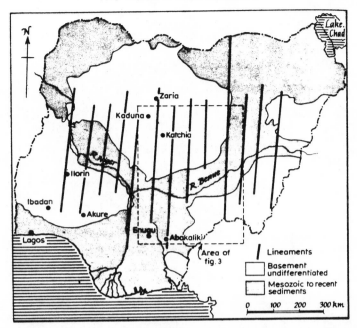

Figure 42.1 *One of a number of sets of long lineaments in Nigeria believed to have been caused by crustal failures over 500 million years ago*

NASA's LANDSAT spacecraft. Most geologists will be surprised that there should be such a long, parallel, closely and regularly spaced system, and that all the lineaments should be so uniformly straight over such lengths. The spacing is of the order of the thickness of the Earth's crust. In the field only vertical movements are detectable along these lineaments (fuller details will be published by us in the February 1977 issue of the *Transactions of the Institute of Mining and Metallurgy*). Two other less continuous sets cross these meridional lineaments at 45°. Unlike the first set these show horizontal shearing, displacing the first set by up to several kilometres in places. These directions show the type of failure pattern that would be caused by a horizontal principle compressive stress in an east-west direction (σ_1 in Figure 42.3).

But what huge stress could cause such a regular pattern of straight failures, each hundreds of kilometres long? Perhaps the near-meridional orientation is a clue and the force may have been related to the Earth's spin.

Figure 42.2 *You need sharp eyes to spot the Nigerian lineaments (arrowed) on this LANDSAT picture. They are part of the set shown in Figure 42.1*

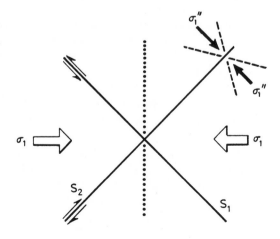

A cosmic near-miss?

As a tentative explanation we suggest the cause as a change of the Earth's speed of rotation due to a near miss by, or an oblique impact with, another large cosmic body, with consequent inertial effects tending to wrinkle the crust. It must be remembered that the Earth's crust is relatively thin – of the order of one hundredth of the radius. The crust instead of bending failed along the lines of the incipient wrinkles. It now shows zones of brittle and ductile failure along the meridional lineaments. The ductile zones, produced where the constraining pressure was greater, were presumably originally deeper in the crust beneath the zone of brittle failure, and have been exhumed by erosion. We suspect that the formation of the meridional system represents one geological episode, and that the shearing may have occurred contemporaneously in terms of geological time.

Dating the dramatic episode that caused it all is difficult. Although the failures show through a surface layer of Mesozoic rocks, and would therefore at first sight appear younger, these formations seem less disturbed along the lineaments than rocks of the old underlying Precambrian basement which is exposed in other parts of Nigeria. It is likely that the fractures were formed in the basement rocks and were later propagated up through the younger sediments by seismic events, Earth tides, or some other phenomenon. The failure zones have provided easy paths for upward movements of bodies of igneous rock, and the lineaments now show as connecting lines of intruded granitic rocks associated with tin deposits, volcanic rocks, hot brine springs, and base metal deposits (Figure 42.4). Granites, rising in molten form from depth through other rocks because of their relatively low density, can contain valuable metals. In a late stage of cooling, these may stream out in a transport system of volatile fluids, settling in fractures to form veins, or reacting with the invaded rock, to form

Figure 42.3 *(opposite) Diagram of the angular pattern of the old fractures in Nigeria. Dotted line represents features shown in Figure 42.1, and σ_1 the principal compressing force likely to have caused them. After first-order shearing (e.g. S_2) there will be a component of σ_1 across the failure (σ_1'') which may generate a system of secondary fractures*

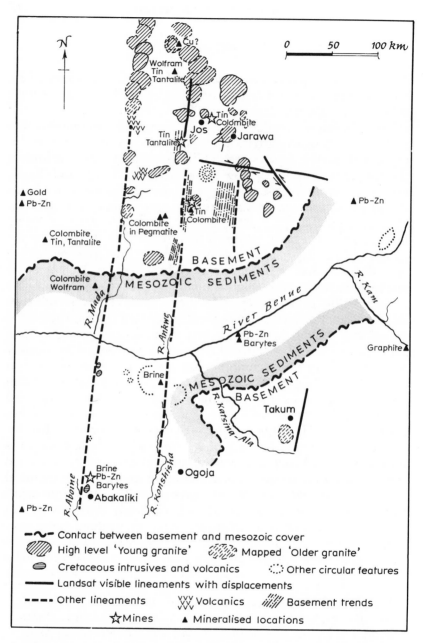

Figure 42.4 *Enlargement of the boxed area of Figure 42.1, showing how the old crustal failures were penetrated by rising granites. A portion of only one shear failure is shown*

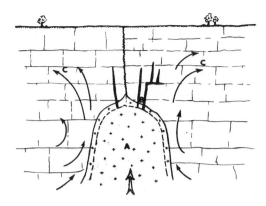

Figure 42.5 *How a hot granite (A) intrudes other rocks forming a hard outer shell that bottles up volatiles. These later stream out through fractures (B) filling them with veins of tin, tungsten, copper etc. The hot granite also converts water and brines in sediment fracture systems, which scavenge dispersed base metals and deposit them in the vicinity (C)*

ore deposits in the vicinity. The hot granite, a potential source of thermal energy, can also act as a heat pump, convecting water and brines from marine sediments (see arrows in Figure 42.5). These can scavenge base metals from the sediments and concentrate them in zones above or adjacent to the intrusion. Thus the lineaments seem to make good strategic prospecting targets for a variety of minerals and thermal energy. They can also play an intimate role in the formation of sedimentary basins with "drape folds" and arched up folds arranged *en echelon* that are of considerable interest to oil exploration teams as potential oil traps.

For some 15 years the photogeology group at Imperial College has been investigating smaller fracture traces showing on air photographs, and using them to resolve regional and local anomalous stress patterns to help detect these types of mineral deposits and potential petroleum reservoir structures. It is now fascinating to find that at least in some areas we can see the larger overall framework into which they fit. Consider the situation of a crustal shear such as S_2 in Figure 42.3. After failure there may be movement along the failure surface, but there will now be a component of the σ_1 force normal to the surface (σ_1'' on Figure 42.3). This stress may generate a second order of failures which, in turn, may cause a third order of smaller failures, and so on. However, the fractures belonging to the lesser orders may be small enough to exist in individual rock types, and the angle between the shear failures now differs in various situations. There are other forms of secondary failures but we have been able to cope with these using a combination of human interpretation, computer, and optical image analyses. Deflections in regional stress trajectories deduced from air photographs are now seen to coincide with these visible traces of deep horizontal shear failure showing on pictures from satellites. Within the overall pattern we have used statistical treatment to detect the local mechanical effects of intrusive rocks, local folding, and settlement of rocks over resistant features – but one always feels that there is more that could be inferred from the chaos of data. The satellite pictures will now give us vital parts of the jigsaw puzzles.

From experience with such types of studies we are surprised to find that the major meridional lineaments in Nigeria remain parallel and straight in the south where they approach the African coast. It seems that they could predate the break when South America separated from Africa and drifted westward. Reported dating studies of intrusive bodies of igneous rock now seen to be along the lineaments show that they reach back to 540 ± 20 million years ago, compared with the approximately 180-million-year age of the break-up of the former supercontinent. There are also some intrusive bodies along the lineaments which show this latter date, indicating disturbance of these old fractures at the time of break-up. We have examined some images of the eastern part of Brazil, believed to have previously fitted alongside Africa, and find a similar style of a major fracturing including some unusual arcuate major fractures which we are still studying. These parallel systems of old crustal failures may be more widespread than has been realised – there are subtle but definite indications that the tin-bearing granites in Cornwall are associated with a north-west trending set showing through the surface cover of sedimentary rocks.

If our guess is correct, we now have a measure of orientation change of some old shield rocks. Thus Nigeria is only 8° from its position relative to north 520 million or more years ago, but the orientation of Brazil has changed about 52°.

10 February, 1977

43

Planet Earth – a view from above

CHRISTINE SUTTON AND DAVID SMITH

Using space techniques first developed to look beyond Earth, scientists will soon know more about the dynamics of the ever-changing planet on which we live.

The Earth is only one of nine planets that orbit the Sun, and during the past two decades man has developed the technology to send spacecraft to observe and explore the other worlds in the Solar System. The US space agency NASA has played a large role in exploiting this new dimension of discovery, with missions to Jupiter, Saturn and beyond. But, dazzled by pictures of volcanoes on some distant moon and grand canyons on another planet, it is easy to forget that NASA also explores planet Earth with the help of that same space technology. Among a variety of activities, NASA supports studies of the Earth's dynamics – particularly the motions of the surface layer of this planet that is very much "alive".

NASA's contribution to a programme of research that also involves the US Geological Survey, the National Oceanic and Atmospheric Administration, and the Defense Mapping Agency, is masterminded from the Goddard Space Flight Center in Maryland, where David Smith is Project Scientist for the Crustal Dynamics Project. A major goal of the joint programme, which was initiated by the Earthquake Hazard Reduction Act of 1977, is to improve understanding of earthquakes, ultimately with a view to reducing the hazards these cataclysmic motions of the Earth's surface present, and possibly, to predicting their occurrences.

The basic data for theories of the dynamics of the Earth concern the movements of the crust. The deceptively simple underlying principle is to measure repeatedly the distances between points on the Earth's surface, and study their variation with time. These

Figure 43.1   *The LAGEOS satellite comprises 426 reflectors which return laser pulses to their points of origin on Earth (Credit: NASA)*

measurements can then be compared with the movements that models of the Earth's crust suggest – in particular with the predictions of plate tectonics: the theory that, put simply, regards the Earth's surface as consisting of a number of rigid plates that ride round on a more plastic layer below. Earthquakes generally occur at the boundaries between plates, often when the edges of two plates that are at first interlocked suddenly slip past each other, releasing vast amounts of stored energy.

The two techniques

Space technology plays a vital role in the science of geodesy, the large-scale measurement of the Earth's shape. Measurements made along the ground by conventional techniques are limited by how far surveyors can see, so large distances can be measured only in many stages, and errors can easily accumulate. But two different techniques – laser ranging and very-long-baseline interferometry (VLBI), both of which owe their heritage to space research – can measure distances from 100 km to the diameter of the Earth (12 700 km) in "one go", and with an accuracy of around 5 cm. Since the mid-1960s, NASA has built up experience in the use of these two techniques which now form the basis of the "field work" for the Crustal Dynamics Project.

Laser ranging involves directing a laser beam at a reflector in space and measuring the time it takes for pulses of light to travel

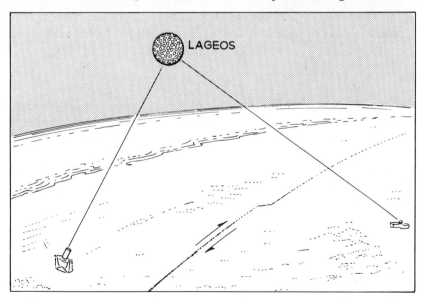

Figure 43.2 *By timing the return journeys from the LAGEOS satellite from different points, scientists can measure the relative distances*

Figure 43.3   *An early version of the mobile laser ranging
station, deployed in the Otay Mountain, near San Diego in
California*

there from Earth and back. This gives the distance from the laser
source to the reflector. If the measurement is repeated from several
points on the Earth's surface, and the orbit of the reflector is
accurately known, then by comparing the different times, the
relative distances between the points on Earth can be calculated.
Several satellites have carried reflectors to allow laser ranging, and
Soviet and American astronauts left reflectors on the Moon so that
Earth-bound scientists could make accurate measurements of its
distance.

However, 1976 saw the launch of NASA's Laser Geodynamics
Satellite (LAGEOS), the first designed solely for laser ranging. This
411-kg satellite consists of an aluminium "golf ball", 60 cm in
diameter, with a brass core, which is covered with 426 "corner-
cube" reflectors that will return any optical signal back to where it
started. Initially, the scientists at Goddard concentrated on
making sure that they knew precisely the satellite's orbit, 5800 km

above the Earth, and on developing the ground-based systems that transmit and receive the laser signals. The collection of geodetic data then began in earnest in 1979.

At locations as far apart as the Goddard centre at Greenbelt, Maryland in the eastern US; the McDonald Observatory in Texas; on the island of Maui, Hawaii; and at sites operated for NASA by the Smithsonian Astrophysical Observatory in Natal, Brazil, and in the Orroral Valley in Australia, lasers will beam up pulses to LAGEOS and other satellites for reflection back to the Earth. Apart from such fixed lasers, NASA has developed mobile laser-ranging systems mounted on trucks that enable members of the project to collect data at a large number of locations, scattered around the globe. Mobile systems are presently set up on Tutuila Island in American Samoa; at Yaragadee in Western Australia; near the Haystack Observatory north of Boston, Massachusetts; at Ford Davis, Texas; at the Owens Valley Radio Observatory and at Goldstone, both in California. In addition the University of Texas has developed a Transportable Laser Ranging Station (TLRS) for NASA. This system, the first of a new generation of highly compact systems, is essentially contained in one small van and can be driven, or even flown, quickly to almost any location. The TLRS will be deployed and operated in areas such as California and South America where a large number of sites need to be visited rapidly.

The complementary technique of VLBI belongs to radio-astronomy. This involves using a number of radio telescopes to view the same distant object. For the geodetic measurements the telescopes look beyond our Galaxy at distant emitters of radio waves, such as quasars – strong radio sources that are among the furthest objects that can be observed from Earth. By bringing together the signals each telescope receives from a particular source, scientists can check how far out of step – or phase – the waves are on arrival at different points on the Earth. The sets of waves produce interference patterns that vary according to this difference in phase. If the telescopes were exactly equidistant from the radio source, the waves would be in phase; the interference pattern reveals the relative positions of the telescopes (Figure 43.4).

As with laser ranging, VLBI can be operated from mobile as well as fixed sites, although mobile VLBI exists only in the US where there is one system, others being planned for the next few years. Mobile laser systems close to VLBI sites, such as the Haystack and

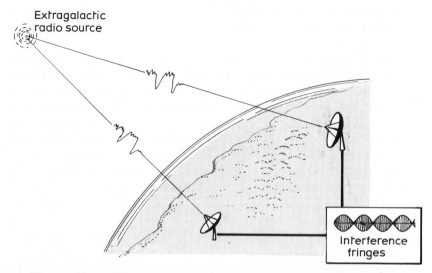

Figure 43.4   *The interference pattern between radio waves, from an extragalactic source, received by two antennas gives an accurate measure of the distance between them*

Owens Valley observatories, will allow NASA to compare the two techniques. Taken together, the two systems should eventually criss-cross the globe in an intricate network of measurements (Figure 43.4).

Smith outlines five major objectives for the Crustal Dynamics Project. The main priority concerns the San Andreas Fault – the boundary between the Pacific and North American Plates – and California in general. The Pacific Plate is moving north-west relative to the North American Plate, which is moving south-east, and points on opposite sides of the San Andreas Fault are slowly moving past each other. This fault makes earthquake research particularly important in the US. Says Smith, "It's a social and political problem, as well as scientifically interesting. The US would certainly like to alleviate the next earthquake; a 1906-type earthquake could do immense damage." Besides, the San Andreas Fault is easily accessible – you can almost literally stand with a foot on either side.

So NASA, in cooperation with the US Geological Survey and the American scientific community, is striving to make precise measurements of the Earth's surface throughout the western US, but principally in California, which is "cracked" in many places by the various faults. One of the big unknowns, according to Smith, is what is happening in a large region around the San Andreas Fault – for 100 to 200 km on either side. The crust is generally thought to deform before a big earthquake, and if this could be observed it might give clues to understanding the "big event". But no one can say when this will be. The aim, meanwhile, is to look for signs of motion, or even the lack of it where it might be expected. As Smith says, if one does see motion, that is not necessarily bad, for the crust is not storing the strain. But in the region around Los Angeles and San Francisco, measurements already indicate that there the two sides of the fault are locked together, so that strain must be accumulating. In the San Francisco earthquake of 1906 nearly 7 m of slippage occurred, and presumably an equivalent amount of strain would have to build up for a similar-sized earthquake to take place in the same region.

Since 1972, two laser stations, one in San Diego and one in Quincy (Figure 45.5), on opposite sides of the San Andreas Fault, have measured the 900 km between them at approximately two-yearly intervals. The distance appears to have decreased by about 60 cm in this time – that is about 8 cm/year – which is about the right amount, considering the uncertainty in these numbers, and in the right direction, according to plate tectonics. But this, as Smith points out, is only a measure of the *total* movement – it doesn't reveal *where* the movement is taking place, and so does not help to solve the earthquake problem directly. Smith would like to be able to lace the San Andreas Fault and the surrounding region, criss-crossing the boundary with measurements from mobile VLBI and laser stations on as many as 30 sites all the way along the fault, from north of San Francisco down into Mexico. The distance between San Diego and La Paz – on the same plate – should not show much movement; San Diego to Mazatlan – across the fault – should change by 6–8 cm/year.

A second priority for the Crustal Geodynamics Project is to put together a global picture of the present tectonic plate movement. Plate tectonics makes predictions about the relative motions of plates and hence the changes in distances between points on the plates. For example, the distance between Washington, on the North American Plate, and Hawaii on the Pacific Plate should increase at a rate of 1.3 cm/year. Conversely, Hawaii and Perth (on the Australian Plate) should be moving closer together by

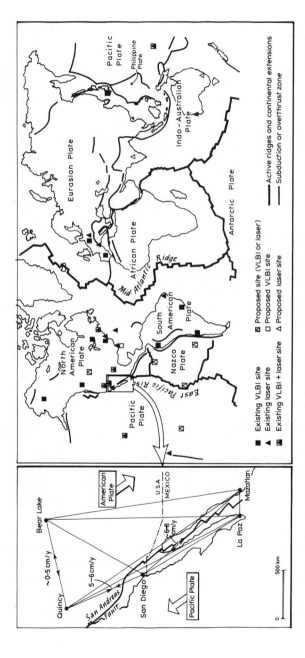

Figure 43.5 *Sites at points around the globe will eventually monitor movements between all the Earth's tectonic plates by accurately measuring the distances between the locations*

some 6.4 cm/year. By studying changes in these distances and those linking other plates over the next 6 to 7 years, Smith hopes that the project will be able to find out just how good a theory plate tectonic is.

Different plates and boundaries

Certainly a more detailed look reveals that the plates are not the rigid blocks a simplistic tectonic theory pictures them to be. Earthquakes *do* occur away from plate boundaries – for example in China, where earthquakes have been felt 1000 km away from the nearest boundary. So a third objective of the Crustal Dynamics Project is to measure the deformation within single plates, while ensuring that different types of plates are studied. For example, earthquakes occur in the eastern US which is part of an old continental plate that has hardly changed for the past 200 million years – since the Atlantic opened. Smith wants to find out to what level this region is stable. Does it change by millimetres or centimetres, say, over a period of several years? In contrast, in the case of oceanic crust, as in the Pacific Plate, scientists believe that enormous chunks of the Earth's surface are moving quite rapidly, at least at their edges. But are they distorting? Such plates are not very thick – less than 30 km – but may be 10 000 km across.

The different types of plate boundary are equally important in the picture of plate tectonics. At some boundaries plates simply slide past each other, as in California at the San Andreas Fault. At other boundaries new crust is created as mantle wells up from below the crust and spreads out. Finally, crust is destroyed at subduction zones, where one plate dives beneath another.

Very active boundaries, such as those along the west coasts of Alaska and South America, should be particularly interesting. By measuring between sites on Easter Island and on mainland South America the project should reveal just how fast the Nazca Plate is moving under South America. Measurements between Easter Island and Hawaii should give information about the rate of sea-floor spreading across the mid-ocean ridge – the East Pacific Rise – that is the boundary between the Nazca and Pacific plates. This is probably the fastest-spreading plate boundary, at some 16 cm/year.

A fifth objective of the NASA programme is to study changes in the rotation of the Earth, the length of day, and motion of the

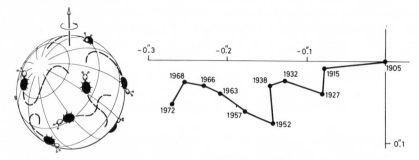

Figure 43.6   *(Left) Beetles crawling over the globe mimic the dynamical effects of moving continents, one of the forces that change the Earth's rotation and the position of the pole (right)*

polar axis. The Earth's rotation can vary with time as changes occur in the distribution of mass around the planet. These changes arise, for instance, from atmospheric motions, earthquakes, plate movements, or processes that take place over a much longer time-scale, such as mountain building. Movement of mass around the Earth does not alter the total angular momentum of the Earth-atmosphere system, but it can change the relative amounts each part contributes. If, for example, some portion of the Earth's crust moves north while maintaining the same angular velocity, it will change the solid Earth's contribution to the total angular momentum, and so alter the length of day (Figure 43.6). The slow deformation before an earthquake might even be reflected in as sensitive a parameter as the Earth's rotation.

Since the end of the past century when regular measurement of the Earth's rotation began, the Earth's spin axis has apparently drifted, while also moving circularly, in a general direction about 90°W (Figure 43.6). This motion, generally called the secular drift, is quite possibly tied up with plate movements and their associated redistribution of mass. Of particular interest over the short term, however, is the relationship between the deformation that earthquakes produce, and the movement of the pole, which is theoretically anticipated to be as much as several tens of centimetres for very large earthquakes.

Smith describes the Crustal Dynamics programme as "very ambitious" and says that it is pushing technology – for example our ability to measure large distances to about 1 part in 10^8, and our knowledge of the behaviour of satellite orbits – a long way.

The two techniques being used represent an "ideal pair" that must continually check each other. Some 20 per cent of the measurements in North America will be made by lasers *and* by VLBI and Smith is "sure that there will be a lot of differences in the first few years". "We don't expect too much now," he says. "The measurements are more important over the long term because then we shall see the changes that occur over the intervening years." Presently planned to take until 1986, at a cost of some $25 million per year – NASA's total budget request for 1981 is $5.7 billion – the programme will "provide an intriguing insight into a number of possibilities, and will ask more questions", according to Smith. He envisages an extension of the programme for some 10–15 years, with data analysis being kept up simultaneously with measurements. An ambitious programme indeed, and one that should provide deeper insights into the dynamics of planet Earth, albeit in a less spectacular style than the public usually associates with NASA's planetary missions.

16 October, 1980

Subject Index

Name Index